IEEE Recommended Practice for Energy Conservation and Cost-Effective Planning in Industrial Facilities

Published by
The Institute of Electrical and Electronics Engineers, Inc

Distributed in cooperation with
Wiley-Interscience, a division of John Wiley & Sons, Inc

IEEE Recommended Practice for Energy Conservation and Cost-Effective Planning in Industrial Facilities

Sponsor

Industrial Plants Power Systems Committee
of the
IEEE Industry Applications Society

ISBN 0-471-82037-7

Library of Congress Catalog Number 84-81026

© Copyright 1984 by

The Institute of Electrical and Electronics Engineers, Inc
345 East 47th Street, New York, NY 10017 USA

November 5, 1984   *SH09472*

Foreword

(This Foreword is not a part of IEEE Std 739-1984, IEEE Recommended Practice for Energy Conservation and Cost-Effective Planning in Industrial Facilities.)

This recommended practice is the result of a six year effort by members of the Industrial Plants Power Systems Energy Subcommittee of the IEEE Industry Applications Society.

The purpose of the IEEE Bronze Book is to promote the concepts and practices of electrical energy conservation in the engineering community. This information was not addressed exclusively by any text available at the time, but rather was distributed among many sources. It has been approved by the IEEE Standards Board as an IEEE standards document to provide current information and recommended practices for energy conservation and cost-effective planning in industrial facilities.

This publication known as the Bronze Book is one of the present IEEE Color Book series sponsored by the IAS Industrial Power Systems Department. This IEEE Recommended Practice will complement and supplement the following color books prepared by the IEEE Industrial and Commercial Power Systems Department:

Recommended Practice for Electric Power Distribution for Industrial Plants (IEEE Red Book), IEEE Std 141-1976.

Recommended Practice for Grounding of Industrial and Commercial Power Systems (IEEE Green Book), ANSI/IEEE Std 142-1982.

Recommended Practice for Electric Power Systems in Commercial Buildings (IEEE Gray Book), ANSI/IEEE Std 241-1983.

Recommended Practice for Protection and Coordination of Industrial and Commercial Power Systems (IEEE Buff Book), IEEE Std 242-1975.

Recommended Practice for Industrial and Commercial Power System Analysis (IEEE Brown Book), ANSI/IEEE Std 399-1980.

Recommended Practice for Emergency and Standby Power for Industrial and Commercial Applications (IEEE Orange Book), ANSI/IEEE Std 446-1980.

Recommended Practice for the Design of Reliable Industrial and Commercial Power Systems (IEEE Gold Book), ANSI/IEEE Std 493-1980.

Comments, corrections, and suggestions for the next revision of this publication are encouraged and should be submitted to the
Secretary
IEEE Standards Board
345 East 47th Street
New York, NY 10017

Working group members and other contributors for this recommended practice were:

Carl Becker, *Chairman*

Norman Blake	H. L. Harkins
Kao Chen	Arthur M. Killin
Melvin H. Chiogioji	John R. Linders
Charles N. Claar	Terry McGowan
Irving Fishman	Gene Montague
Ernest M. Freegard	William Moylan
Daniel L. Goldberg	Wayne L. Stebbins

Rudy Verderber

At the time of approval of this standard the members of the Energy Subcommittee were as follows:

Carl Becker, *Chairman*

James H. Beall	C. Grant Keough
James W. Beard	Arthur M. Killin
K. W. Carrick	John R. Linders
Kao Chen	Alfred B. Marden
Melvin H. Chiogioji	Thomas Mason
Charles N. Claar	Bal K. Mathur
Daniel L. Goldberg	W. J. Moylan
L. E. Griffith	Hasamettin Ovunc
H. L. Harkins	Milton D. Robinson
Thomas D. Higgins	Eugene R. Smith
Len Ilgen	Wayne L. Stebbins
James Iverson	Myron Zucker

When the IEEE Standards Board approved this standard on June 23, 1983, it had the following membership:

James H. Beall, *Chairman*   **Edward Chelotti,** *Vice Chairman*

Sava I. Sherr, *Secretary*

J. J. Archambault	Donald N. Heirman	John P. Riganati
John T. Boettger	Irvin N. Howell, Jr	Frank L. Rose
J. V. Bonucchi	Joseph L. Koepfinger*	Robert W. Seelbach
Rene Castenschiold	Irving Kolodny	Jay A. Stewart
Edward J. Cohen	George Knomos	Clifford O. Swanson
Len S. Corey	R. F. Lawrence	Robert E. Weiler
Donald C. Fleckenstein	John E. May	W. B. Wilkens
Jay Forster	Donald T. Michael*	Charles J. Wylie

*Member emeritus

Energy Conservation and Cost-Effective Planning in Industrial Facilities

Working Group Members and Contributors

Carl E. Becker, *Working Group Chairman*

Richard H. McFadden, *Chairman Publications*, Industrial Power Systems Department

Section 1 — Introduction: Charles N. Claar and Daniel L. Goldberg, *Cochairmen*

Section 2 — Organizing for Energy Conservation: Melvin H. Chiogioji and Arthur M. Killin, *Cochairmen*

Section 3 — Translating Energy Into Cost: Carl E. Becker and Charles N. Claar, *Cochairmen*

Section 4 — Load Management: Kao Chen, *Chairman*

Section 5 — Conservation Considerations in Electrical Machines and Equipment: John R. Linders, *Chairman*

Section 6 — Metering and Measurement: Wayne L. Stebbins, *Chairman*

Section 7 — Energy Conservation in Lighting Systems: Kao Chen and Terry McGowan, *Cochairmen*

Section 8 — Cogeneration: H. L. Harkins, *Chairman*

Contents

FIGURES

1. Introduction

1.1 General Discussion. IEEE Std 739-1984, IEEE Recommended Practice for Energy Conservation and Cost-Effective Planning in Industrial Facilities commonly known as the IEEE Bronze Book, is published by the Institute of Electrical and Electronics Engineers (IEEE) to provide a recommended practice for electrical energy conservation in industrial facilities. Most of the material herein is also applicable to commercial facilities. It has been prepared by engineers and designers on the Energy Subcommittee of the IEEE Industrial Power Plant Systems Committee (IPPS) of the IEEE with the assistance of the Production and Application of Light Committee (PAL).

This recommended practice will probably be of greatest value to the power-oriented engineer with some design or operation experience with industrial and commercial facilities. It can be an aid, however, to engineers and designers at all levels of experience. It should be considered a guide and reference rather than a detailed manual, to be supplemented by the many excellent publications available.

There will be an overlap with other fields of engineering, particularly mechanical and architectural, in dealing with building systems which are covered in this recommended practice.

When references are made to codes, standards, laws, and regulations, it is essential that the document referred to or current authentic interpretation be consulted. Such material changes frequently in contrast to conventional engineering information, and it is impractical to include complete current regulations or detailed interpretations in a text of this size.

1.2 Conservation. Energy conservation as treated in this recommended practice deals with engineering, design, applications, utilization, and to some extent the operation and maintenance of electric power systems to provide for the optimal use of electrical energy. Optimal in this case refers to the design or modification of a system to use minimum overall energy where the potential or real energy savings are justified on an economic or cost benefit basis. Optimization also involves factors such as comfort, healthful working conditions, the practical aspects of productivity, aesthetic acceptability of the space, and public relations.

1.3 Methodology. The methodology of conservation involves the use of the management tool commonly referred to as energy management. This has been elevated to a separate *discipline* or specialty. It often is even more confined by being emphasized within the conventional disciplines — electrical or mecahnical. The term *energy* engineer will refer to the engineer performing duties in the energy area, whether full or part time.

1.4 Energy Management. Energy management, the more inclusive term for energy conservation, involves the following professions and fields:

(1) Engineering
(2) Management, organization
(3) Economics
(4) Financial analysis
(5) Operations research (system analysis)
(6) Public relations (selling conservation)
(7) Environmental engineering

Some of the tools that are dealt with here are:

(1) Meters and measurement
(2) Demand and energy limiters
(3) Highly efficient energy devices
(4) Control systems — building management systems

While this recommended practice emphasizes electrical energy conserva-

tion, other forms of energy shall be treated. Section 8 deals with the combined efficient use of electrical and thermal energy (steam) in a highly efficient cost-effective manner. The text points out that the reduction of energy in one area can actually have a negative effect on the heat balance. For example, the reduction of thermostat temperature setting in large buildings in winter can cause excessive operation of the air-conditioning chillers.

1.5 Periodicals. The energy engineer needs to keep up-to-date on new technologies in this rapidly expanding field. The following periodicals offer new engineering information on energy conservation.

1.5.1 *Spectrum*, the basic monthly publication of the IEEE, covers all aspects of electrical and electronic engineering including conservation. This publication also contains references to IEEE books and other publications, technical meetings and conferences, IEEE groups, societies, and committee activities, abstracts of papers and publications of the IEEE and other organizations, and other material essential to the professional advancement of the electrical engineer.

1.5.2 *The Transactions of the Industry Applications Society* (IAS) contain papers presented to conferences, a number of which deal with energy conservation. All members of the IAS receive this publication.

1.5.3 Publications which specifically are oriented toward energy conservation include:

Energy User News.[1] A weekly news publication in tabloid format which covers the entire field of energy conservation

---

[1] Fairchild Publications, 7 East 12th Street, New York, NY 10003.

with emphasis on keeping the reader informed of new technologies. It is quite open in discussing the effectiveness or failures of systems. A legislative and regulatory scoreboard is included.

NTIS — National Technical Information Service of the US Department of Commerce[2] is an abstract newsletter published weekly. Each year it contains abstracts of hundreds of articles relating to energy listed by subject. A purchase form for obtaining many of the articles is included with each issue.

1.5.4 The Association of Energy Engineers publishes several documents including the bimonthly *Energy Engineering* and the quarterly *Energy Economics Policy and Management.* In addition, the Association and the US Department of Energy sponsored a conference in 1981. The papers are available in the *Proceedings of the Fourth World Energy Engineering Congress.*[3]

1.5.5 Some other periodicals of a more general nature which are heavily involved with electrical energy conservation are:

Electrical Construction and Maintenance, 1221 Avenue of the Americas, New York, NY 10020.

Electrical Consultant, One River Road, Cob, CT 06807.

LD & A, Illuminating Engineering Society, 345 East 47th Street, New York, NY 10017.

Plant Engineering, 1301 South Grove Avenue, Barrington, IL 60010.

Power, 1221 Avenue of the Americas, New York, NY 10020.

---

[2] 5285 Port Royal Road, Springfield, VA 22161.

[3] PO Box 14227, Atlanta, GA 30324.

Power Engineering, 1301 South Grove Avenue, Barrington, IL 60010.

Electric Light and Power, 1301 South Grove Avenue, Barrington, IL 60010.

Specifying Engineer, 51 South Wabash Avenue, Chicago, IL 60603.

1.5.6 Two technical periodicals which contain material of interest are:

Energy Management is directed at management in business and industry. It is published bi-monthly by Penton, 614 Superior Avenue West, Cleveland, OH 44113.

Plant Energy Management is published bi-monthly by Walker-Davis Publication, Inc, 2500 Office Center, Willow Grove, PA 19090.

1.5.7 It should be noted that new periodicals can be expected in the expanding field of Energy. Some of the publications are available at no cost to qualified individuals engaged in energy conservation work.

1.6 Standards and Recommended Practices. A number of organizations in addition to the National Fire Protection Association (NFPA) publish documents which affect electrical design. Adherence to these documents can be written into design specifications.

The American National Standards Institute (ANSI) coordinates the review of proposed standards among all interested affiliated societies and organizations to ensure a consensus approval. It is in effect the *clearing house* in the US for technical standards of all types.

Underwriters' Laboratories, Inc (UL), is a nonprofit organization, operating laboratories for investigation of materials and products, especially electrical appliances and equipment with respect to hazards affecting life and property.

The Edison Electric Institute (EEI) and the Electric Power Research Institute (EPRI) represent the investor-owned utilities and publish extensively. The following handbooks are available through EEI:[4]

Electrical Heating and Cooling Handbook.

A Planning Guide for Architects and Engineers

A Planning Guide for Hotels and Motels Electric Space-Conditioning.

Industrial and Commercial Power Distribution

Industrial and Commercial Lighting

The National Electrical Manufacturer's Association (NEMA) represents equipment manufacturers. Their publications standardize the manufacture of and provide testing and operating standards for electrical equipment. The design engineer should be aware of any NEMA standard which might affect the application of any equipment that he specifies.

The IEEE publishes several hundred electrical standards relating to safety, measurements, equipment testing, application, and maintenance. The following three publications are general in nature and are important for the preparation of plans.

ANSI/IEEE Std 100-1984, IEEE Standard Dictionary of Electrical and Electronics, Terms.

ANSI/IEEE Std 315-1975, IEEE Standard Graphic Symbols for Electrical and Electronics Diagrams.

ANSI Y32.9-1972, American National Standard Graphic Symbols for Electrical Wiring and Layout Diagrams Used in Architecture and Building Construction.

The Electric Energy Association (EEA) publishes specifications and informative pamphlets for buildings with electric space conditioning.

The Building Officials and Code Administrations International, Inc (BOCA), is an organization which promulgates the BOCA building construction model codes. It also provides educational services in code administration and enforcement. Many governmental bodies have mandated its codes as the governing construction code. The BOCA Basic Energy Conservation Code 11981 is a compilation of model energy conservation requirements from the BOCA Basic Building Code applicable in 1981.[5]

1.7 Industry Application Society (IAS). The IAS is one of 31 IEEE groups and societies which specialize in various technical areas of electrical engineering. Each group or society conducts meetings and publishes papers on developments within its specialized area. The IAS presently encompasses 23 technical committees covering electrical engineering in specific areas (petroleum and chemical industry, cement industry, glass industry, industrial and commercial power systems and others). Papers of interest to electrical engineers and designers involved in the field covered by this text are, for the most part, contained in the Transactions of the Industry Applications Society of the IEEE.

---

[4] 111 19th Street, Washington, DC 20036.

[5] BOCA is located at 17926 Halsted Street, Hanewood, IL 60430.

1.8 IEEE Publications. The IEEE publishes several standards similar to the IEEE Bronze Book prepared by the Industrial Power Systems Department of the IEEE Industry Applications Society.[6]

[1] ANSI/IEEE Std 142-1982, IEEE Recommended Practice for Grounding of Industrial and Commercial Power Systems (IEEE Green Book).

[2] ANSI/IEEE Std 241-1983, IEEE Recommended Practice for in Commercial Buildings (IEEE Gray Book).

[3] ANSI/IEEE Std 399-1980, IEEE Recommended Practice for Power System Analysis (IEEE Brown Book)

[4] ANSI/IEEE Std 446-1980, IEEE Recommended Practice for Emergency and Standby Power Systems for Industrial and Commercial Applications (IEEE Orange Book).

[5] ANSI/IEEE Std 493-1980, IEEE Recommended Practice for the Design of Reliable Industrial and Commercial Power Systems (IEEE Gold Book).

[6] IEEE Std 141-1976, IEEE Recommended Practice for Electric Power Distribution for Industrial Plants (IEEE Red Book).

[7] IEEE Std 242-1975, IEEE Recommended Practice for Protection and Coordination of Industrial and Commercial Power Systems (IEEE Buff Book).

[8] IEEE J2112-1-1973, Protection Fundamentals for Low-Voltage Electrical Distribution Systems in Commercial Buildings.

1.9 Governmental Regulatory Agencies

1.9.1 Actions of the United States government regulatory agencies are embodied in the *Federal Register*.[7] Often the rules have sufficient legal complexity that interpretations are best left to experts in the field. The agencies making such rules shall operate within the scope of laws enacted by Congress.

1.9.2 The Department of Energy (DOE) sponsors experimental or demonstration projects through grants and other incentives. Reports are issued on such programs as, for example, the fluorescent lamp solid-state ballast under grant to Lawrence Berkley Laboratories of the University of California.

1.9.3 Individual states have their own Departments of Energy. Agencies such as the DOE are able to exert a powerful influence on these local departments through law or by the ability to withhold funds for areas deemed not in compliance with federal regulations. At the present time ASHRAE 90-75 and ASHRAE 90-100[8] are accepted as a basis for design of new buildings and existing buildings. While some states have accepted these standards almost verbatim, other states, such as New York, have issued their own equivalents. It is important to recognize which rulings are based on law in the locality under consideration.

---

[6] These documents are available from the Sales Department, IEEE Service Center, 445 Hoes Lane, Piscataway, NJ 08854.

[7] This daily publication is available from the US Government Printing Office, Washington, DC 20402.

[8] These documents are available from ASHRAE, Publication Sales, 1791 Tullie Circle, NE, Atlanta, GA 30329.

1.9.4 ASHRAE 90-75 and ASHRAE 90-100 were developed by the American Society of Heating, Refrigeration and Air-Conditioning Engineers (ASHRAE) in conjunction with members of the Illuminating Engineering Society (IES) to provide an energy budget. The production of this recommended practice was encouraged by various governmental agencies to fill a need for a standard developed on a consensus basis by voluntary professional organizations knowledgeable in energy conservation. These documents have legal status only where local authorities have incorporated their recommendations.

1.9.5 Where generation (except for emergency and standby purposes is involved) and especially where the resale and redistribution of energy for use by others are concerned, regulations of the Federal Energy Regulatory Commission (FERC) may apply. PURPA, the Public Utilities Regulatory Act of 1978, now under review by the courts, provides guidelines for the sale and resale of energy and for exemption from FERC rulings which are applicable to utilities. Industries employing cogeneration or generation not requiring fossil fuels fall under PURPA jurisdiction. These industries are exempt from certain FERC regulations. They also have certain advantages in their relations with utility companies.

1.9.6 Public Service or Utility Commissions *govern* the actions of the local utilities. They enact rules and establish rates, after public hearings, responsive to the needs of the utilities and their customers and to the pressures of governmental bodies as described above. During recent years, these Commissions have recognized energy conservation needs by establishing time of day, incremental, cogeneration and similar rates

intended to minimize energy and peak-demand power usage. The effects of such actions are expressed in the rates and rules of the local utilities.

1.9.7 Utilities usually provide energy conservation services in addition to their more conventional advice on rates and conditions of service. Some utilities will provide energy audits and some may be required to check the energy *budget* before providing service. However, the advice they provide is usually more general than that of the consultant or in-house engineer.

1.10 Keeping Informed. The following suggestions offer ways to remain current in the abundance of information being generated in the energy field.

(1) Read the aforementioned periodicals. Some are dedicated to energy only, while others can be scanned for energy articles.

(2) Attend energy-oriented conferences and courses, hundreds of which are given throughout the country. The courses are often conducted for one or two days and may deal with the subject in general or concentrate on special aspects. Sessions covering the subject are presented frequently by the organizations listed in 1.10 (3).

(3) Become active in the energy committees of societies such as the IEEE, IES, AIA (American Institute of Architects) and ASHRAE on a local or national basis.

(4) Read professional books and handbooks dealing with conservation. These will often be advertised or listed in the periodicals referred to above.

(5) Read advertising literature, which, although often biased, can provide an excellent guide to new equipment, its application and the basis of justification.

(6) If you are a manager, establish the programs, described in the following sections, utilize the services of consultants, and attend energy seminars specifically designed for the manager, executive architect, financial planner etc.

1.11 Professional Activities. If an engineer is to practice publicly, professional registration is usually essential. Many organizations require such registration for certain levels of engineering. Regulatory agencies require that designs for public and commercial buildings be prepared under the jurisdiction of state-licensed professional architects or engineers. Information on such registration may be obtained from the appropriate state agency or from the local chapter of the National Society of Professional Engineers.

To facilitate obtaining registration, in different states under the reciprocity rule, a national professional certificate is issued by the National Bureau of Engineering Registration to engineers who obtained their home-state license by examination. All engineering graduates are encouraged to start on the path to full registration by taking the engineers-in-training examination as soon after graduation as possible. The final written examination in the field of specialization is usually conducted after four years of progressive professional experience.

1.12 Coordination with Other Disciplines. The *energy engineer* in his work will often span several of the conventional disciplines and will often deal with other disciplines.

1.12.1 The energy engineer is concerned with professional associates such as the architect, the mechanical engineer, the structural engineer, and, where underground services are involved, the civil engineer. He is also concerned with the builder and the building owner or operator who, as clients, may take an active interest in the design or redesign. The engineer will work directly with the architect. He shall cooperate with the safety engineer, fire-protection engineer, the environmental engineer, and a host of other concerned people, such as interior decorators, all of whom may have a say in the ultimate design or modification for conservation.

In performing conservation design, it is essential, at the outset, to prepare a checklist of all the design stages that have to be considered. Major items include staging, clearances, access to the site, and review by others. It is important to note that certain conservation work may appear in several chapters of the general construction specifications. Building control systems, even if electronically based, may appear in the mechanical section or in a separate section. Many organizations utilize the specification format of the Construction Specification Institute (CSI) where the energy conservation work of the electrical-mechanical engineering will usually fall in Chapter 15, Chapter 16 (electrical) and (if included) a Chapter 17 (special building control systems). For example, furnishing and connecting of chiller electric motors may be covered in the mechanical section of the specifications. For administrative purposes, the work may be divided into a number of contracts, some of which may be awarded to contractors of different disciplines. Among items with which the designer will be concerned are: preliminary estimates, final cost estimates, plans, or drawings, specifications (which are the written presentation of the work), materials, manuals, factory inspections, laboratory tests, and temporary power. He may well be involved in providing information on how the proposed conservation items affect financial justification of the project in terms of owning and operating costs, amortization, return on investment, and related items.

1.12.2 The energy engineer is encouraged to proceed with care in making changes which affect people. A test installation is often justified in determining the acceptability of an installation. Human reactions to changed environment often cannot be modified without considering the other physical factors. Some typical pitfalls are listed below:

(1) Are lighting color changes acceptable and compatible?

(2) Are workers uncomfortable in the changed ventilation conditions?

(3) Is task lighting producing a gloomy environment?

(4) Is machinery operating at reduced power because of voltage reduction drop-out in a brownout?

(5) Is the area uncomfortably humid because of loss of reheat?

(6) Is the building computer debugged?

(7) Is a perfectly *good* lighting redesign producing a rash of headache and eye complaints?

(8) Are the building tenants sharing in energy money savings?

The experienced energy engineer will have encountered most of these conditions, which are covered in this text, and be able to avoid the possible negative side of energy conservation. Today, a host of new tools described in the following sections are available to ease the work of the engineer in preparing a well

conceived, energy-conserving design.

1.13 Text Organization. The subsequent sections cover energy conservation in the following sequence of subjects:

1.13.1 Fuels. In addition to introducing conservation, this section covers the affect of fuel cost on electric energy cost.

1.13.2 Organizing for Energy Conservation. The proper organization and management commitment are prerequisites to a successful energy conservation effort.

1.13.3 Economics. This section provides several models for evaluating energy alternatives and motor and transformer loss evaluation. Electric rate structures are also explained.

1.13.4 Load Management. The use of demand control can reduce electric bills but not necessarily energy use. This section looks at demand control techniques but, more importantly, discusses energy control techniques.

1.13.5 Electrical Machines and Equipment. Proper application of equipment along with correct design and maintenance of the electrical system and its components are key factors in energy conservation and system efficiency.

1.13.6 Metering. Proper measurement techniques ensure a viable and continually effective energy conservation program.

1.13.7 Lighting. New devices and proper application of existing equipment can provide significant energy savings.

1.13.8 Cogeneration. The concurrent use of steam for process (and comfort) heating, and electric generation provides a very efficient process, provided that specific requirements are met.

Design details are too intricate for discussion in this text. The reader is, therefore, advised to make extensive use of design manuals and other sources.

1.14 Fuels. All known fuels or energy sources have a life cycle that begins with the development of a means to effectively manufacture and utilize the fuel itself. The energy source then grows in usage to a point of maximum production at which the usage begins to decline. This decline is caused by many factors, including obsolescence and depletion of supply. Today's energy sources are divided into three major categories — fossil, solar, and nuclear fuels.

1.14.1 Fossil. The three most common fossil fuels are coal, oil, and natural gas. Their cleanliness, heat content, and availability will affect their use and electrical costs in the future, so these aspects are covered in this section.

(1) Coal is generally acknowledged as the United States' most plentiful fossil fuel resource. At current consumption rates the supply should last at least 100 years. The Btu content and quality of coal varies from region to region, and it is not uniformly available throughout the country. Problems affecting the use and availability of coal are: mining, including site restoration and water pollution; transportation; air pollution; and lack of a coherent governmental coal-burning policy that encourages the use of this resource. However, industry and government have recently increased funding for coal research. Some new techniques are being explored for the use of coal; coal burning technology should improve in the coming years.

One advantage of coal as an energy source is that the industry is unregulated. This fact makes coal costs subject, to some extent, to laws of supply and demand. However, government health and safety mining regulations, in addition to inflationary pressures and the energy situation in general, have caused coal prices to soar. Coal prices should track inflation under the present conditions.

(2) Natural gas is the cleanest burning fossil fuel, but it has the least proven reserves. Many parts of the country have experienced shortages of natural gas. Outright curtailments have often forced the shutdown of plants, schools, and commercial establishments.

The industry was highly regulated; however, the National Gas Policy Act (NGPA) initiated the process of deregulation. Interstate regulation of natural gas by the Federal government was accomplished by fixing the price at the well head. Intrastate regulation was done by state utility commissions because natural gas companies generally had exclusive distribution and selling rights in a given geographical area.

Natural gas presents few air pollution problems and is probably the least capital intensive of the fossil fuels. Lack of availability remains its chief disadvantage, although deregulation should encourage exploration and the use of wells that were previously thought to be uneconomical to operate. However, US reserves will probably be depleted in two decades at current consumption rates.

The NGPA will cause gas prices to keep pace with the rate of inflation. Deregulation is scheduled for completion by 1987. For industrial gas customers, one negative aspect of the NGPA is that interstate pipeline and distribution companies shall price the gas "incrementally;" that is, the price of new gas shall be passed on to the large user as opposed to all users of gas. In summary, gas prices should rise as the rate of inflation, but deregulation will adversely affect the price if the supply fails to keep pace with demand.

(3) Oil is the second cleanest burning fuel, and it is not very capital intensive

when compared to coal. Because of these two factors, it has distinct advantages as an industrial and commercial fuel. Availability could be a future problem. While the United States imports oil there should be few problems with availability. The price of domestic oil was regulated by the Federal government and is now in the process of deregulation, while the world price of oil is unregulated.

The problems caused by the oil embargo of 1973 are well documented. In summary, the price of foreign oil has accelerated at a rate which has no relationship to the laws of supply and demand. The US demand for oil has outstripped our domestic supply. Without foreign oil, we would deplete our reserves in approximately two decades. The US is confronted, therefore, with a problem concerning the use of oil now and in the immediate future. Oil prices will rise faster than inflation.

1.14.2 Solar. Solar energy encompasses the use of the sun's radiation to produce electricity or as a heating and cooling source. Solar energy has gained attention because of governmental and general public pressure, but it has not been an economically viable energy resource. More research and expanded use of solar energy, even in situations where it is uneconomical, should improve the technology and lower capital costs.

The five primary types of solar energy include: photovoltaic, thermal, wind, hydropower, and biomass. Photovoltaic involves the direct conversion of the sun's rays into electricity. Thermal conversion uses the sun's rays to create heat which is either used directly or to create steam for electric generation. Wind is generally used to provide shaft horsepower to a generator. Biomass primarily involves trash as a heat source for a steam-driven turbine generator.

(1) Photovoltaics are probably the most interesting means of converting solar energy into electricity. There are currently three drawbacks to photovoltaic that make widespread use doubtful. First, they are only 15% efficient while costing 10 to 100 times more per kilowatthour than other methods. Second, a very large area is required for collection of the energy. A 132 MW demonstration plant in Phoenix, Arizona, uses 640 acres or an equivalent of a one-mile by one-mile plot. Third, the system can only generate electricity on sunny days.

A very intriguing proposal uses an earth orbiting solar station to collect the light and then converts the energy to microwaves beamed to an earth-based receiving station. This concept allows 24 h generation without any negative effects of weather. While much more efficient, the system requires significant amounts of land for the receiving station and is much more expensive than earth-based photovoltaic. The system is apparently safe, but the affects of the microwave beam on radio, TV, and wildlife have not been determined.

(2) Thermal conversion has not proven economical for generating electricity. However, heating and cooling systems are now marketed that compete favorably with other energy sources on a five to ten-year payback basis. A central collector has been proposed for use in generating steam for a turbine driven generator. However, the land requirement is large at over 150 acres per electric megawatt output, and output is restricted to daylight hours.

(3) Wind conversion has been successfully used to generate small amounts of power. The drawbacks with wind genera-

tion include noise, radio and TV interference, very fast tip speeds (60–80 ft/s) and size limitation. The NASA device in Sandusky, Ohio, has 50 ft diameter blades to power a 225 kW generator. The NASA device operates when the wind speed is between 7 and 30 m/h, which is typical. Higher electrical output is being generated in windmill farms containing many mills.

(4) Most electricity generated today by hydropower is by high head dams, large scale operations that only large investors can afford to finance. However, low head dams are becoming more attractive as a means to generate electricity. The use of low head dams could improve the availability of electricity on a local basis.

(5) Biomass is being used in conjunction with other fuels, due to low heat content. This source may begin to supply small amounts of power for municipalities and industrial complexes. As with coal, there are many pollution and handling problems.

1.14.3 Nuclear. There are three nuclear conversion techniques — fission, breeder, and fusion reactors. While this is the cleanest and most economical source of power, regulations and political pressures make future dependence on this source questionable.

Fission conversion has proven economical for three to four decades. However, the disposal of nuclear waste and eventual plant decommissioning methods have not received universal concurrence. Furthermore, the known uranium reserve would only supply energy for 100 years to a nation totally reliant on fission power.

The breeder reactor can multiply the nuclear supply by a factor of ten by recycling nuclear waste. This concept has been demonstrated on a large scale, but development in the United States has been virtually at a standstill.

The fusion concept has been sustained on a laboratory level but full development is not anticipated for decades. Fusion could ultimately provide a virtually unlimited supply of energy at a very low cost without pollution. However, the infancy of fusion technology makes this source inappropriate for consideration at this time.

1.14.4 Conclusion. From all indications, the cost of electricity will increase at a rate at least equal to that of annual inflation. Coal will probably be the primary source of fuel in the United States. Petroleum and gas will probably increase in price at a rate faster than electricity, but the relative costs and rates will differ in various areas.

2. Organizing for Energy Conservation

2.1 Introduction. To understand the energy consumption patterns in the industrial sector, it is important to understand the applications of energy in industrial processes. Industrial energy applications are grouped into six major types.

2.1.1 Types of Industrial Energy Applications

(1) *Space Conditioning.* Energy used directly for heating or cooling an area for comfort conditioning without first being converted to steam or hot water.

(2) *Boiler Fuel.* This is subdivided into space conditioning and process energy, depending on how the steam or hot water from the boiler is utilized.

(3) *Direct Process Heat.* Energy used to heat the product being processed, for example, for kilns, reheat furnaces, etc, excluding energy used in boiler steam or hot water.

(4) *Feedstock.* Fuel used as an ingredient in the process.

(5) *Lighting*

(6) *Mechanical Drive.* Motors used for pumps, crushers, grinders, production lines, etc possess this mechanical drive.

Energy savings can occur by improving the efficiency of energy conversion processes, by recycling the waste energy or, by reusing the waste materials.

Any process requires a certain minimum consumption of energy. Energy (or equipment) additions beyond this minimum require an evaluation of the incremental cost of more efficient equipment or techniques versus the resulting energy savings or costs. Some of the more intensive users of industrial energy, including chemicals, paper, and petroleum refining, have long found it competitively advantageous to design for energy conservation. However, other less energy intensive industries have not found the use of energy recovery equipment economically advantageous.

Two economic incentives exist for the development of an energy conservation program on a plant-by-plant basis: (1) the savings realized by reducing energy use, and (2) preventing economic losses by minimizing the probability of fuel supply curtailment.

With increasing costs of energy, every manager can benefit by seeking to reduce energy consumption.

2.1.2 Energy Saving Methods. The ways in which energy can be saved are grouped into four general categories:

(1) *Housekeeping Measures.* Basically better maintenance and operation. Such measures include shutting off unused equipment, improving electricity demand management, reducing winter temperature settings, turning off lights, and eliminating steam, compressed air and heat leaks.

(2) *Equipment and Process Modifications.* These can either be applied to existing equipment (retrofitting) or be incorporated in the design of new equipment. Examples include the use of more durable or more efficient components, the implementation of novel, more efficient design concepts, or the replacement of a process with one using less energy.

(3) *Better Utilization of Equipment.* This can be achieved by carefully examining the production processes, schedules, and operating practices. Typically, industrial plants are multi-unit, multiproduct installations which evolved as a series of independent operations with minimum consideration of overall plant energy efficiency. Improvements in plant efficiency can be achieved through proper sequencing of process operations; rearranging schedules to utilize process equipment for continuous periods of operation to minimize losses associated with start-up, scheduling process operations during off-peak periods to level electrical energy demand, and conserving the use of energy during peak demand periods.

(4) *Reduction of Losses in the Building Shell.* Reduction in heat loss is achieved by adding insulation, closing doors, reducing exhaust, and by utilizing process heat, etc.

Management shall provide effective personnel motivation, planning, and administration to achieve meaningful energy savings. The establishment of a formalized energy management responsibility is highly desirable to give the effort both the focus and direction required. An energy management function gives the line manager the tools to get the job done. Line managers need to know their energy use and costs, the future energy supply availability and its cost, problems or opportunities of energy situations, and those alternative solutions worth pursuing.

The following sections attempt to identify and describe the steps needed to implement an effective energy management program. The engineer or plant manager can be a catalyst in developing a solid, functional energy conservation program.

2.2 Organizing the Program. At any level of the corporate structure, an individual should understand the incentives and motivations of top management. An alert engineer will become aware of the hidden lines of authority and aware of the key persons who make decisions. These key persons shall be convinced of the positive merits of an energy conservation program before it can be successful. Therefore, it is the engineer's job to see that the proper facts are presented through the proper channels to convince top management that they should make the energy commitment. Sometimes the informal organizational structure is an effective tool. The five critical factors in organizing an effective energy conservation program are:

(1) Obtain top management energy commitment. This is a formally communicated, financially supported dedication to reducing energy consumption while maintaining or improving the functioning of a facility. This commitment shall be active and clearly and visibly communicated to all levels of the organization in terms of words and actions by top management.

(2) Obtain people commitment. People in all levels of the organization should be involved in the program. Ideas should be encouraged with rewards for cooperation. People should be shown why their help is needed, and a team approach should evolve. The most successfully planned program can be devastated by a single person trying to subvert the program.

(3) Set up a communication channel. Use the channel to report to the organization the results, high achievers, rewards, etc. Use the channel to advertise the program and to encourage cooperation.

(4) Change or modify the organization to give authority and commensurate responsibility for the conservation effort and develop a program.

(5) Set up a means to monitor and control the program.

An organizational plan, using the above criteria, should then be developed for both implementing and monitoring specific energy conservation programs. This plan should also: define the responsibilities of the energy conservation coordinator or committee; describe an effective communication system between coordinator and major divisions, departments, and employees; establish an energy accounting and monitoring system; and provide the means for educating and motivating employees. Realistic goals should be established that are specific in both amount and time. The goals should be in terms of energy and not cost, while taking into consideration the effect of diminishing returns.

The energy team should at least consist of representatives from each major facility or department; engineering, operations, and the union or labor representative. Other members might include personnel from purchasing, accounting, and finance. The team should have authority adequate to investigate prevailing energy supply and demand situations and to implement policy recommendations. Their tasks should be assigned a priority consistent with the current or potential importance of energy problems in the company operation. They should utilize those talents within the firm in making the analysis and implementing the programs.

2.3 Surveying Energy Uses and Losses. The first step for the energy team is to determine by an energy audit the amount of energy that enters and leaves a plant. This determination will probably be an approximation at first, but the accuracy should improve with experience. The audit consists of a survey and appraisal of energy and utility sytems at various levels of detail.

There is a direct relationship between the extent of data collection and consolidation, and the subsequent evaluation of energy conservation opportunities. While an insufficient data base may prevent the identification of several energy-saving opportunities, too extensive a baseline survey may prove unnecessary and wasteful by diverting funds and time from more rewarding conservation opportunities. While the best results can be obtained by conducting a thorough, comprehensive survey and analyzing all site energy and utility systems, time and budgetary constraints may impose limitations on the extent of survey and appraisal for various site energy systems.

Certain items shall be covered. Power bills shall be reviewed considering the following quantities, when available: kilowatthours, kilovarhours, kilowatt demand, kilovoltampere demand, power factor, days in billing cycle, reading date, weather conditions, and production factors affecting power consumption. The plant should be toured to discover such items as: unnecessary operation of equipment, unnecessary high levels of lighting, no one assigned to turn off unnecessary lights and equipment, improper thermostat setting, unnecessary or excess ventilation, and obvious waste. A comprehensive tour will also involve acquisition of nameplate data, motor and lighting loads, and other system details.

The amount of attention given will depend on the type of audit. A walk-through audit will require only a general notation of the performance while a comprehensive audit will require specific information. Six categories for energy use should be analyzed as the tour is conducted. These six categories are as follows:

(1) Lighting: interior, exterior, natural, and artificial

(2) The Heating, Ventilating and Air Conditioning (HVAC) system and the heating and cooling effects of conduction, convection, and radiation

(3) Motors and drives

(4) Processes

(5) Other electrical equipment — transformers, contactors, conductors, switchgear, etc)

(6) Building shell, in terms of infiltration, insulation, and transmission.

Within each of these categories, five basic conservation concepts shall be considered:

(a) General housekeeping — turn off unused equipment, close doors, redirect heat or light, etc)

(b) Process or physical changes — move the desk closer to the window

(c) Better use of equipment or purchase of new equipment — run two batches instead of one

(d) Obvious misapplication of equipment — [150 fc (1600 lx) in a storage area or a 10-ton crane to move a 100 lb carton]

(e) Maintenance — change filters, clean lights, lubricate motors and drives

2.3.1 Lighting. The first item that gets attention is lighting because of its visibility. Before making changes read the standards on lighting. Key efforts in analyzing and considering lighting from a conservation standpoint are noted below. It is wise to make checks both day and night.

(1) Is the light intensity sufficient for the task? [15-40 fc (160-450 lx) for halls; 70-100 fc (750-1100 lx) for detailed work.]

(2) Is the fixture proper for directing the light where it is needed?

(3) Is the reflection good?

(4) Is the color right for the task?

(5) Is the fixture too high or too low?

(6) Can task lighting be used effectively?

(7) Is good use being made of available natural light?

(8) Can desks or machines be grouped by task light required?

(9) Are lights and fixtures cleaned periodically?

(10) Are lights turned off when not in use?

(11) How many fixtures can be turned off by a single switch?

(12) Who turns lights off?

(13) Who uses the space and how often?

(14) Can a different, lower wattage bulb be used in the fixture?

(15) Do the surfaces reflect or absorb light?

(16) Are the lights strategically located?

(17) Does the fixture location cause glare?

(18) Can light be used to heat?

(19) Can more efficient light sources be used?

(20) Can timers or photocells be effectively used?

2.3.2 Heating, Ventilating and Air Conditioning (HVAC). Key factors in evaluating and better utilizing the HVAC system are as follows:

(1) Are there obstructions in the ventilating system?

  (a) Do filters, radiator fins, or coils need cleaning?

  (b) Are ducts, dampers, or passages and screens clogged?

(2) Is the wrong amount of air being supplied at various times?

  (a) Are dampers stuck?

  (b) Is exhaust or intake volume too high or too low?

  (c) Are all dampers functioning in the most efficient manner?

(3) Can the system exhaust only the area needing ventilation?

(4) Can the system intake only the amount required?

(5) Can air be recycled rather than exhausted?

(6) Can the intake or exhaust be closed when the facility is unoccupied?

(7) Can the system be turned off at night?

(8) Is the temperature right for area's use? — 40 °F- 50 °F can be good for storage

(9) Can temperature setback be used effectively?

(10) Will variable speed drive be more efficient?

(11) How many fixtures can be turned off by a single switch?

(12) Is solar energy being effectively utilized?

  (a) Light but minimum heat in summer

  (b) Light and heat in winter

(13) Can waste heat be used?

(14) Are belts tight?

(15) Are pulleys and drives properly maintained and lubricated?

(16) Is the refrigerant proper?

(17) Is waste heat utilized?

(18) Can heat be redirected?

(19) Is the proper system being used?

(20) Is there too much or too little ventilation?

(21) Can the natural environment be used more effectively?

(22) Are doors, windows, or other openings letting out valuable heat?

(23) Can weatherstrip, caulk, or other leaks be repaired?

(24) Can additional insulation be justified?

(25) Can all hooded exhaust systems have their own air supply or can they be used as part of the exhaust requirement for the building?

(26) Is the blower cycled or run continuously?

2.3.3 Motors and Drives. Because motors use nearly 70% of the electrical energy consumed in the US, they provide great opportunity for reducing energy waste. The following questions point to very common kinds of waste:

(1) Does the motor match the load?

(2) Can the motor be stopped and then restarted rather than idled?

(3) Is the motorized process needed at all? Can it be done by hand?

(4) Who lubricates the motor and associated drives? Is this done at the proper interval?

(5) Can motor heat be recycled?

(6) What type of drive is used? Is it the most efficient?

(7) What is the voltage and is it balanced?

(8) Can the motor be cleaned to lower heat buildup?

(9) How is load adjusted?

(10) Will two (or more) motors in tandem work better?

(11) Is the motor well maintained and in good condition? Are there any electrical leaks to ground? Is it in a wet environment?

(12) Who turns the motor off and on? How often?

(13) How efficient is the motor?

2.3.4 Processes. Normally, processes rely heavily on motors, but there are other electrical parts. Process heating is probably the most common nonmotor electric process load. The following questions point to areas where improved efficiencies can be made:

(1) Can equipment or processes be grouped together to eliminate the transportation of the equipment or material in process?

(2) Is the temperature too high?

(3) Does heat escape? Can insulation be used effectively?

(4) Can the heated energy be recirculated for comfort, process heat or cogeneration? Can it be exhausted for summer comfort?

(5) Is preheat required?

(6) Can the process be staged or interlocked?

(7) Is the product heated, cooled and then reheated again? Therefore, is a continuous process more appropriate?

(8) Can the processes be lined up for more effective use of equipment?

(9) Are the drives, bearings, etc lubricated?

(10) Can the conveyor system be eliminated or modified?

(11) Can hot areas be isolated from cold areas?

(12) Is one large or many small motors better?

(13) What equipment can be turned off at night?

(14) Would two or three shifts be more efficient?

(15) Is any equipment kept idling rather than shut off when in hold or waiting?

(16) Are screens cleaned, dampers checked for proper operation of pollution controls, and are they maintained at proper intervals, etc?

(17) Is compressed air made in two or three stages? Is a hold tank being used and is pressure too high?

(18) Is process water too hot?

(19) Can fluid be recirculated?

(20) Is the fluid cooled too much?

(21) Is the hot water heater close to hot water use?

(22) Is the process exhaust higher than required for safety or quality, or both?

(23) Is hot and cold piping insulated where appropriate?

(24) Is temperature controlled so that only necessary heat is added?

(25) Is heat supplied or added at the point of use or is it transmitted some distance?

2.3.5 Other Electrical Equipment. There is a significant amount of electrical equipment that is taken for granted or rarely noticed. The list below asks some key questions regarding efficiency:

(1) Are the transformers required?

(2) Is the transformer too hot? (Good heat source.)

(3) Can the transformer be turned off when not in use?

(4) Are wiring connections tight? Improper voltage, unbalanced voltage, and excess heat can result from a bad connection.

(5) Can heat from the switchgear room be utilized? Remove this heat during the summer.

(6) Are voltage taps in the proper position?

(7) Are heaters applied properly?

(8) Can heaters be turned off at times?

(9) Are contactors in good working order?

(10) Is equipment properly bonded and grounded?

(11) Are conductors sized properly for the load?

(12) Is the power factor too low?

2.3.6 Building Environmental Shell. The list below is applicable to electrically heated buildings and also where other energy sources are used.

(1) Is allowance made for transition from cold to hot areas and vice versa with an air curtain or vestibule?

(2) Can a wind screen keep air infiltration down?

(3) Is the proper level of insulation applied?

(4) Can full advantage of solar heat be taken? Remove effect in summer.

(5) Is automatic door closing appropriate?

(6) Can covered loading and unloading areas be utilized to keep heat in?

(7) It is possible to caulk, weatherstrip, glaze, or close-off windows?

(8) Can double or triple pane glass be used?

(9) Is a small positive pressure used to keep out drafts?

(10) Can areas be staged in progressively cooler or warmer requirements?

(11) Would an air screen, radiant heater, etc be more effective?

(12) Are dock seals used on overhead doors?

2.3.7 Overall Considerations. All conservation changes shall be made while considering the overall plant. An electrically heated building will use the heat from lighting fixtures in the winter, so reduced lighting may just transfer the heat requirement to baseboard heaters. While this section provides many areas for improved efficiency, a detailed analysis and engineering design by a competent consultant will be required. The final design will include the consideration of proper codes (National Electrical Code, Occupational Safety and Health Administration Codes, National Fire Protection Association Codes, and Environmental Protection Agency Codes, etc).

2.3.8 Energy Balance. An important step in evaluating opportunities is the development and analysis of major energy balances. An energy balance is a little more complicated than an energy survey and is based on measurement and calculation.

A basic energy balance is shown in Fig 1. Energy is input to a system which

Fig 1
Energy Balance

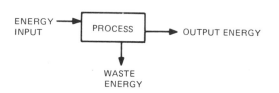

produces a product. There is a certain amount of energy absorbed in the process which can be called energy output. The difference between the energy input and the energy output is waste. Energy balances should be developed on each process to define, in detail, energy input, raw materials, utilities, energy consumed in waste disposal, energy credit for byproducts, net energy charges to the product, and energy dissipated or wasted.

All process energy balances should be analyzed in depth. Various questions should be asked and answered, including:

(1) Can waste heat be recovered to generate steam or to heat water or a raw material?

(2) Can a process step be eliminated or modified in some way to reduce energy use?

(3) Can an alternate raw material with lower process energy requirements be used?

(4) Is there a way to improve yield?

(5) Is there justification for replacing old equipment with new equipment requiring less energy or replacing an obsolete, inefficient process with a new and different process using less energy?

The energy survey and balances identify energy wasting situations and differentiate between those that can be corrected by maintenance and operations actions and those that require capital expenditures. The former can be corrected in a short time and the results are almost immediate — savings with little effort or delay. The latter will require some investment dollars and delivery time for materials and equipment.

The survey can be conducted by an individual, a team, or an outside consultant. The work can be segregated according to plant areas or departments.

However, those conducting the survey shall be aware of the importance of finding waste, determining the cost involved, and reducing the energy costs. The initial survey shall be followed up periodically with additional surveys to ensure that waste is curbed and new problems are avoided.

2.4 Energy Conservation Opportunities. A key element of the energy management process is the identification and analysis of energy conservation opportunities (ECOs). These opportunities involve the previously listed spectrum of activities, from simply changing procedures or turning off lights to new long-range technologies.

The equipment used in manufacturing plays another key role in the energy conservation effort. By understanding the relative energy consumption by equipment type, one can determine the opportunities that exist for decreasing this consumption.

After developing the energy balance and listing all of the energy conservation projects, each project should be evaluated for implementation using the following procedure:

(1) Calculate annual energy savings for each project.

(2) Project future energy costs and calculate annual dollar savings.

(3) Estimate the cost of the project, including both capital and expense items.

(4) Evaluate investment merit of projects using measures such as return on investment, etc.

(5) Assign priorities to projects.

(6) Select conservation projects for implementation and request capital authorization.

(7) Implement authorized projects.

Section 2.3 describes examples of electrical energy conservation opportunities. Application of a specific ECO requires careful evaluation to determine if its use is appropriate because there is the possibility that an ECO could be counterproductive. For instance, the proposed change may cause existing equipment to exceed operating limits.

2.5 Energy Monitoring and Accounting. An effective energy conservation program requires feedback on performance. Since company-wide energy conservation is relatively new at most companies, the best techniques are not yet apparent. However, the ensuing paragraphs describe some guidelines that have evolved.

An energy accounting system is essential for control and evaluation of an energy conservation program. Different uses and applications require different accounting and measuring techniques. In many cases, periodic or irregular fuel deliveries or meter readings will not provide adequate information to determine the variation of energy use with daily, weekly, or monthly production cycles, or with seasonal meteorological changes. In other cases, energy in the form of fuels, electricity, and process steam will be used in two or more different major applications in the plant. In some energy applications, such as in combustion processes, measurement or control, or both, at each point of application

may be necessary to ensure efficient use of energy. Measuring methods that best describe the process are thus necessary to control, evaluate, and manage efforts to conserve energy.

A number of very useful methods to maintain energy conservation performance in plants and offices worldwide are described in [1].[9] Other established effective methods are given in [3], [4], [5], and [6]. In engineering new facilities, potential conservation is monitored using either the percent reduction energy rate method or the design energy savings report. For existing facilities, [1] states that energy conservation is monitored using the activity method, the energy-rate method, the variable energy-rate method, or tracking charts.

2.5.1 Percent Reduction Energy Rate Method. This reduction is determined by comparing the project's product energy rate, expressed in Btu/lb with the existing plant's product energy rate. An example is shown in Table 1.

The product energy requirements consist of all energy supplied minus credit for all usable energy exported. The requirements include purchased energy such as gas, oil, and electricity, plus generated inplant utilities such as steam,

---

[9] The numbers in brackets refer to the numbers of the Bibliography in 2.8.

Table 1
Product Energy Rate

Current production rate = 1950 Btu/lb

Projected production rate = 1480 Btu/lb

$$\text{Percent reduction in product energy rate } = \frac{1950 - 1480}{1950} \cdot 100$$

$$= 24\%$$

refrigeration, compressed air, and cooling water. All energy, purchased or generated has to be expressed in the same units, for example, Btu. Likewise, the manufacturing department is given credit for usable energy which is exported such as condensate and steam.

2.5.2 Design Energy-Savings Report. The design energy-savings report covers an energy-savings idea that is incorporated in the design of a project. The report serves three purposes.

(1) It is a means of information exchange on energy-saving ideas.

(2) It provides an opportunity to monitor energy conservation capital requirements versus energy savings on projects.

(3) It assists in an energy-awareness program in the engineering department.

This report is prepared by a project participant when he incorporates an energy-saving idea or innovation that will reduce energy requirements. The guideline for an energy-saving idea is that it deviates from present design or plant practices for the product line. If it is a new product, the reference point is the pilot plant or project definition report.

A summary report is issued quarterly on the number of energy-saving ideas submitted by each design group. This quarterly report encourages competition between design groups.

2.5.3 Activity Method. The activity method compares the anticipated energy saved with energy purchased. The percent energy savings is based on the annual energy savings, expressed in Btu, compared with the plant's total purchased energy.

The activity method gives a rapid response to conservation results. Equally important, savings are not affected by changes in product energy efficiency as the production rate varies. For these reasons, it is an excellent method for monitoring performance.

A typical quarterly report for a company using the activity method is that shown in Table 2. A similar report can also be prepared expressing the savings in Btu and costs — statistics that are most useful for internal communications and external publicity.

The activity method yields a good visual report on activities that have been completed and planned to save energy. However, it does not reflect the true re-

Table 2
Activity Method Report

XYZ Manufacturing Co Energy-Conservation Results Activity Method (% Energy Savings)					
Plant	3 Qtrs	4th Qtr	Year 0 Total	Planned Activities Year 1	Year
Fulton	2.0	0.2	2.2	2.1	1.0
Grace	1.8	0.1	1.9	0.8	0.4
Nixon	1.5	0.2	1.7	2.3	0.3
St. James	0.8	0.4	1.2	0.1	0
Company Total	1.7	0.3	2.0	1.8	0.4

Table 3
Energy-Rate Method Report

Fulton plant Products manufactured	Year 0 (Base Period)			Year 3			
	Total Production (× 10^6 lb)	Total Energy (× 10^6 Btu)	Base Energy Rate (× 10^6 Btu)	Total Production (× 10^6 lb)	Total Energy (× 10^6 Btu)	Comparison Base Period Energy (× 10^6 Btu)	% Reduction Energy Consumption Rate
	1	2	3	4	5	3 × 4 = 6	7
A	200	10 000	50	300		15 000	
B	10 000	30 000	3	12 000		36 000	
C	2000	20 000	10	3000		30 000	
D	3000	60 000	20	6000		120 000	
	15 200	120 000		21 300		201 000	
Adjustments to base: new products (+) or discontinued products (−) after Year 0 (base).							
E (Year 1)	1000	10 000	10	2000		20 000	
F (Year 2)	1000	5 000	5	1000		5 000	
	2000	15 000	—	3000		25 000	
Total products	17 200	135 000		24 300	208 000	226 000	7.5*
Adjustments for environmental and OSHA	Base				(1000)		
Adjusted grand total	17 200	135 000		24 300	208 000	226 000	8.0*

*% Reduction energy consumption rate = $\dfrac{(\text{Column 6}) - (\text{Column 5})}{(\text{Column 6})} \cdot 100 = (\text{Column 7})$.

duction in energy rate, Btu per pound of products, because the savings are annualized.

2.5.4 Energy-Rate Method. The Chemical Manufacturers Association (CMA) developed the chemical-industry energy-rate method for reporting energy conservation results. An example of the calculations is shown in Table 3. The Year 0 base period rate is calculated using the total pounds of product manufactured in Year 0 and the equivalent purchased energy consumed by the department manufacturing the product. The equivalent energy includes energy consumed by the specific line plus a proportionate share of energy that cannot be assigned (or measured) to any one area.

For any reporting year the percent reduction in energy consumption rate is calculated by comparing the base period energy to the total energy purchased by the plant, excluding feedstock energy. The comparison base period energy for each product is calculated using the base year product energy rate times the weight of the product manufactured in the current year.

The CMA energy-rate method of reporting conservation results compensates for product mix, the addition and deletion of products, and for OSHA and environmental energy requirements. Changes in production rate have a major effect on conservation results because the product energy-rate requirements are made up of fixed and variable energies.

2.5.5 Variable Energy-Rate Method. Many variables need to be considered in a conservation monitoring method, including:

(1) Changes in energy rate as production rate varies. See Figs 2 and 3.

(2) Product energy requirement

**Fig 2
Change in Production**

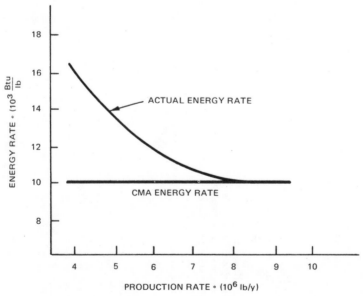

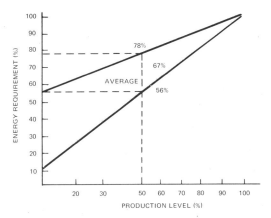

Fig 3
Energy Production

changes resulting from ambient temperature changes during the year.

(3) Changes in raw material quality.

(4) Changes in the plant heat balance that may cause the process to operate turbine drives instead of motor drives, and vice versa.

(5) Minor changes in product quality or specifications for different customers.

(6) A change in the equipment or process to increase output.

The product energy rate versus production-rate curve in Fig 2 was developed for each product considering only production rate and its effect on the energy rate as variables. To calculate the conservation performance for any year, the product production rate for that year is used to establish the base period energy rate from the energy-rate curve of Fig 4. For example, using the CMA energy-rate method for Year 0, the product energy rate is 10 000 Btu/lb. The actual base energy rate is 11 500 Btu/lb.

The 11 500 Btu/lb base energy rate instead of 10 000 is used in calculating the comparison base-period energy rate for product G. This same procedure is then duplicated for each product. This method, although not 100% accurate in compensating for all variables, does compensate for the major variable, production rate, and does prevent wide swings in conservation results as the production rate varies.

2.5.6 Tracking Charts. The tracking charts in Fig 5 can be used to establish trends in energy usage resulting from conservation, maintenance or the lack thereof, and operational techniques. Tracking charts are a speedy current technique used to track department and total plant production, purchased energies, generated utilities, and product energy rates. Also, tracking charts can monitor trends in conservation, energy usage, and potential problem areas.

Two time frames are typically used. Short periods (shift to daily) are used to review performance with shift operating personnel for quick feedback on operational performance. These usually span a month or so. Weekly to monthly period charts provide data that tend to average fluctuations in energy usage, and consideration is given to trends over a longer time. Any analysis of tracking charts shall take into consideration production rate and the effect it has on energy usage.

2.6 Employee Participation. An energy conservation program can be successful only if it arouses and maintains the participative interest of the employees. Employees who participate and who feel themselves partners in the planning and implementation of the program will be more inclined to share pride in the results.

Communicating with employees on the subject of energy can be accomplished in many different ways: face to face discussion, seminars and workshops, distri-

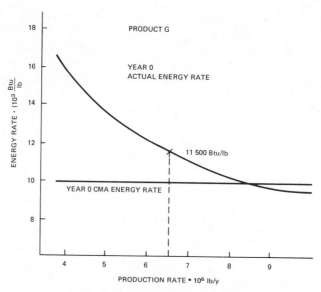

**Fig 4
Actual and CMA Energy Rate
Versus Production Rate**

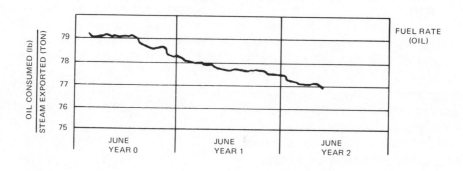

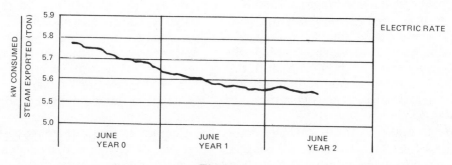

**Fig 5
Tracking Chart**

bution of informative and descriptive literature, slide presentations and moving pictures, and most important of all, sincere practice of conservation on the part of management at all times.

The use of company newsletters, bulletin boards, or posters for pictorializing energy conservation objectives and accomplishments will help impress employees with the importance of such matters. Employee participation can be increased by communicating examples of energy conservation ideas being implemented, photographs of persons who submitted the idea, and information on the savings realized.

Competition between departments, sections or groups within the company in pursuing conservation of energy can also generate enthusiasm among employees. Competitive programs can be initiated among the employees and should be encouraged. Acknowledgment of good ideas and positive reinforcement are keys to this approach. Employee education can take many diverse forms: workshops and training courses for supervisory personnel, articles in the company newsletter, and energy conservation checklists given to each employee.

A clear, concise list of firm *do's* and *don't's* to guide employees in performance of their work can be helpful in achieving energy conservation practices. Such lists should be distributed to all employees whose jobs involve the use or control of energy. The list should be updated as often as necessary. Supervisors should be responsible for seeking adherence to all items on the lists.

2.7 **Summary.** Energy management is a broader concept than energy conservation. In energy management, one is attempting to achieve the same level of productivity with a lower expenditure of energy, an adequate energy supply, and the lowest possible cost.

Energy management responsibility requires a significant reorientation of managers toward energy conservation. Corporate individualism operating under the profit motive in a free marketplace has been a major factor contributing to US economic success. These achievements have also been due, in part, to the industrial manager's dedication to the substitution of inexpensive electric and thermal energy for human labor in the production of goods and services.

We now find that it is necessary for management to accept the reality that the future requires a major reorientation because energy input is no longer going to be inexpensive. To secure energy supplies and to minimize energy costs, industry will have to act in conjunction with its own members, government, the public, and all segments of the energy supply industry.

Achievement of meaningful energy savings in existing processes is a function of management committing itself to do the job by effective personnel motivation, planning, and administration. The establishment of a formalized energy management responsibility is highly desirable to give the effort both the focus and direction required. An energy management function gives the line manager the tools to get the job done. Engineers shall inform upper management regarding energy use and costs, future energy supply cost and availability, and the level of savings possible. The engineer shall then convince management of the need for alternative solutions which should be pursued.

Obtaining participation from key local plant operators is a prime ingredient to energy conservation success since they know where the energy leaks occur and generally are in a position to act fast. However, top management shall give them the monetary and technical resources required.

2.8 Bibliography

[1] DOERR, R. E. Six Ways to Keep Score on Energy Savings, *The Oil and Gas Journal*, May 17, 1976, pp 130-145.

[2] *Energy Efficiency and Electric Motors*, Arthur D. Little, Inc, Springfield, VA National Technical Information Service (Pub PB-259129), 1976.

[3] GATTS, ROBERT; MASSEY, ROBERT; ROBERTSON, JOHN, *Energy Conservation Program Guide for Industry and Commerce* (NBS *Handbook 115*), Washington, DC, US Department of Commerce Sept 1974.

[4] GIBSON, A. A. et al, *Energy Management for Industrial Plants*, British Columbia, Canada, Dec 1978.

[5] MYERS, JOHN, et al. *Energy Consumption in Manufacturing*, Massachusetts, Barlinger Publishing Co, 1974.

[6] NEMA AND NECA, *Total Energy Management*, A Practical Handbook on Energy Conservation and Management, 1979, National Electrical Contractors Association, 2nd ed.

[7] *Site Energy Handbook*, Washington, DC, Energy Research and Development Administration, Oct 1976.

[8] STEBBINS, Wayne, Implementing An Energy Management Program, *Fiber Producer*, Atlanta, GA. W. R. C. Smith Publishing Company, Oct 1980, vol 8, no 5, pp 44-57.

3. Translating Energy into Cost

3.1 Introduction. The use of economic analysis is critical to the conservation program, because the monetary savings can significantly influence management decisions. The engineer shall, therefore, be able to translate his proposal(s) into monetary values (that is, expenditures versus savings). This section covers basic economic concepts and utility rate structures and then addresses the subject of loss evaluation.

The engineer can use the economic basics to develop his energy program and to determine the best choice among alternatives. This determination shall include understanding payback periods, understanding the time value of money, and weighing pertinent costs.

An understanding of electric rate structures is essential in an economic analysis because the monetary savings will accrue from lower electric bills. Rate structures are sufficiently complex to warrant careful consideration. The electric rate is particularly important in demand control (load management) projects.

All electrical equipment has losses. The monetary loss reduction is calculated by evaluating the reduced cost of energy over the life of the equipment.

3.2 Important Concepts in an Economic Analysis. Two or more alternatives are considered in an economic analysis — one may be to do nothing. In any case, an investment is usually evaluated over a period of time. The initial investment is called the *capital cost (investment)*. Since an item is usually worth less as it ages the value is depreciated over its life. A piece of equipment has a *salvage value* when it is retired and sold. In addition to the initial investment, an alternative usually has recurring costs such as maintenance and energy usage. These costs are grouped as *annual costs* and are then put in a form that can be added directly to the capital cost.

The capital cost represents the total expenditure for a physical plant or facility. The capital cost is comprised of two components — direct costs and indirect costs. Direct costs are monetary expenditures that can be directly assigned to the project such as material, labor for design and construction, debugging costs, etc. Indirect costs (overheads) are monetary expenditures that cannot be directly assigned to a project such as taxes, rent, employee benefits, management, corporate offices, etc.

Depreciation is the distribution of a capital cost over the anticipated life of the process or equipment. These depreciation amounts are then used to reduce the value of the capital investment. There are several means of distributing these costs, and the simplest method is *straight-line depreciation*. In straight-line depreciation, the annual depreciation is merely the capital cost divided by the estimated life.

There are two *lives*, the *book life* and the *expected (useful) life*. The book life is the number of years used to financially depreciate the investment. The expected life is the anticipated length of time that the investment will be utilized. Computers tend to become obsolete in only a few years due to rapid advances in technology today. However, computers can, and do, continue to compute long after their technological obsolescence: so the expected life in the case of a computer could be the time until obsolescence.

The *salvage value* is the amount of money that can be returned to the company at the end of the expected life of the capital investment. This salvage value is equal to the anticipated resale value minus any cost associated with (1) the sale of the equipment and (2) its physical disassembly and subsequent removal.

The annual costs include such items as fuel (energy), operation/maintenance, labor, taxes, and other recurring costs.

The fuel cost depends on the amount, quality, rate of use, and the tariff (rate schedule). The maintenance cost includes both routine work and purchase of parts, such as the periodic relining of boilers.

3.3 Time Value of Money

3.3.1 Determining the Cost of Money. A dollar today is worth more than a dollar in the future because today's dollar can generate profit. Most companies use the cost of borrowing and the return on investment to determine the cost of money. The energy engineer should work with the appropriate financial people to determine the cost of money.

Inflation, the rate of price increase, is a very important concept in the time value of money. For example, inflation may direct a decision to buy now or next year. If an item increases in cost by 20% a year and money can be borrowed at 10% a year, one should buy now. However, if the reverse is true, it might be wise to wait a year to purchase. In simplified economic evaluations, inflation is assumed to have equal effects on all alternatives; it should only be included in more complex analyses or if one or several items increase in cost at significantly different rates.

The time value of money is important to the engineer because he shall be able to evaluate alternatives by translating a dollar of expense or investment at various times to equivalent amounts. To accomplish this task, the next section will develop conversion factors that translate future dollars or payments into their present values and vice versa.

3.3.2 Calculating the Time Value of Money. The following list of terms is used in subsequent discussions.

annuity. A series of equal amounts

evaluated at the end of equal time periods (usually one year) for a specified number of periods.

capital investment. The amount of money invested in a project or piece of equipment (this includes labor, material, design, and debugging monies).

compound interest. Interest that is applied to both the accumulated principal and interest. For example, a 12% annual interest rate compounded monthly on $1.00 is $(1.01)^{12}$ or $1.13 which results in a 12.7% effective (simple) annual interest rate.

constant dollars. The worth of a dollar amount in a reference year, including the effect of inflation. Constant dollars are used in economic indicators.

current dollars. The worth of a dollar today.

depreciation. The mathematical distribution of the capital investment over a given period of time which may or may not concur with the estimated useful life of the item.

discount rate. The percentage rate used by a corporation that represents the time value of money for use in economic comparisons.

future worth (value). The value of a sum of money at a future time.

initial cost. Synonymous with capital investment. Capital investment is a more proper and clearly recognized term.

present worth (value). The value of an amount discounted to current dollars using the time value of money.

For the purpose of this section, it is assumed that the investments are made at the beginning of the year and the expenditures are made at the end of the year. Hence, $1.00 invested now at a 10% annual interest rate will grow to $1.10, at the end of the first year, but $1.00 in operating and expense dollars will have a $0.91 present value (1.00 · 1/1.1 = 0.91). The present worth is the value at year zero and the future worth is the value at the end of the nth year. The entire conversion process is based on two basic calculations: the present worth of a single amount and the present worth of an annuity.

The single present worth factor is used to find the present worth of a single future amount. The present worth factor (PWF) is determined by the following equation:

$$PWF = \frac{1}{(1 + i)^n}$$

where
  i = interest or discount rate expressed as a decimal
  n = number of years.

The present worth is calculated by multiplying the future amount by the present worth factor or

$$PW = PWF \cdot FW$$

where
  PW = present worth
  FW = future worth or future amount

For example, suppose that a company has the choice of refurbishing an existing induction heating unit for $10 000 (and replacing it in 5 years for $120 000) or buying a new unit for $100 000. The company's discount rate is 12%. The choice is whether to spend $10 000 now and $120 000 in five years or to spend $100 000 now. To compare the alternatives, the $120 000 future amount is converted to today's dollars.

Alternate A

Present: = $10 000 PW

Future: = $\dfrac{1}{(1 + 0.12)^5} \cdot 120\,000$

    = 68 100 PW

Total: = $78 000

    present worth

Alternate B

Present: = $100 000 PW

Future: = 0

Total: = $100 000

    present worth

In this case there is a clear $22 000 benefit in renovating the unit and paying a larger amount for a new machine in five years. Ignoring the time value of money would have led to a wrong decision.

The present worth of an annuity factor (PAF) converts a series of future uniform payments to a single present worth amount. The uniform payments are made at the conclusion of a series of equal time periods. The mathematical description of this factor is as follows:

$$PAF = \frac{(1 + i)^n - 1}{i(1 + i)^n}$$

   = present worth of an annuity factor

and

$$PW = PAF \cdot AP$$

where

AP = annuity amount

   (or equivalent annual payment)

Suppose the induction heater in the previous example had an energy cost of $20 000 per year but the new unit only cost $15 000 per year to operate.[10] The

---

[10] Operating costs commonly include maintenance, refurbishing, etc, but in this simplified example, only energy is considered.

annuity factor will provide a base of comparison by including the annual energy cost as a single present worth amount. While the net penalty or savings can be used, the actual amounts for each alternative should be used to reduce errors and clarify results.

Alternate A

Current expense: $10 000 PW

Future capital investment: 68 100 PW

Future energy costs =

$$\frac{(1 + 0.12)^5 - 1}{0.12\,(1 + 0.12)^5} \cdot \$20\,000 = 72\,100 \text{ PW}$$

Total: $150 190 PW

Alternate B

Current capital investment $100 000 PW

Future energy costs =

$$\frac{(1 + 0.12)^5 - 1}{0.12\,(1 + 0.12)^5} \cdot \$15\,000 = 54\,000 \text{ PW}$$

Total: $154 000 PW

While the choice is still Alternate A, intangible factors could change the decision since the difference is only 2.7%.

The future worth factor (FWF) converts a single current dollar amount to a future amount. The future worth factor for a single present value is the reciprocal of the single present worth factor or

$$FWF = (1 + i)^n$$
$$FW = FWF \cdot PW$$

The future worth of an annuity factor (FAF) converts an annuity to a single future amount.

$$FAF = \frac{(1 + i)^n - 1}{i}$$

and

$$FW = FAF \cdot AP$$

The uniform annuity factor (UAF) is used to convert a single present amount to a series of equal annual payments. The uniform annuity factor is the reciprocal of the present worth of an annuity factor.

$$UAF = \frac{i(1 + i)^n}{(1 + i)^n - 1}$$

and

$$AP = UAF \cdot PW$$

It is frequently necessary to open a savings account for a future purchase, and the amount saved is called a *sinking fund*. The annuity (sinking fund) required to accumulate some future amount is determined by using the sinking fund annuity factor (SAF). The factor is simply the reciprocal of the future worth annuity factor or

$$SAF = \frac{1}{(1 + i)^n - 1}$$

The preceding factors are summarized in Table 4. However, most studies will use only the present worth of a single future amount and the present worth of an annuity factor. The next section describes the use of these factors in an economic analysis of energy options.

3.4 Economic Models. There are two general means of evaluating energy options: simple break-even analysis and a more complex method called life cycle costing. When there is a large energy savings for a small investment, use of the time value of money may not be required. Furthermore, housekeeping projects with minimal or no cost may not need evaluating at all. For example, the installation of a timer on an exterior light circuit that will significantly reduce *on-time* or a decision to turn off unused equipment may not need an economic evaluation. However, the subsequent effect of these actions should be shown to encourage management and support the energy conservation effort. Life cycle costing is most likely needed when the project costs are large compared to the energy savings or when there are significant future costs. The ensuing sections detail the various modeling methods.

3.4.1 Break-Even Analysis. The break-even methods do not use the time value of money, and they all answer the same question: At what point will I get my money back? Common terms for this model are Simple Payback Analysis, Break-Even Point Analysis, and Minimum Payback Analysis. All of these methods are essentially the same. They all relate the capital investment to the savings. Some methods chart results while others use a not-to-exceed value. The basic mathematical description is

$$\text{Break-even point} = \frac{\text{net capital investment}}{\text{savings/unit}}$$

Many companies have minimum payback requirements to allow expenditures whose break-even point is less than a given amount of time (or other measurement). The break-even point has taken the name of *payback* from this minimum payback usage. Hence, the more common term is payback, and management is more likely to ask for the payback of a particular energy option.

It is important to note that break-even analysis is not restricted to time as a base. Production and energy consumption are also good bases.

When questions concerning future operations are appropriate, a more sophisticated method of analysis is justified particularly if the minimum payback period exceeds several years. A com-

Table 4
Time Value Factors

Symbol	Noun Name	Description	Formula*
AP	Annuity payment	Equal amounts of money at the ends of a number of periods	—
FAF	Future worth annuity factor	Converts an annuity to an equivalent future amount	$FAF = \dfrac{(1+i)^n - 1}{i}$
FW	Future worth	The dollar amount (of an expense or investment) at a specific future time	—
FWF	Future worth factor	Converts a single present amount to an amount at a future point in time	$FWF = (1+i)^n$
PAF	Present worth annuity factor	Converts an annuity to a single present amount	$PAF = \dfrac{(1+i)^n - 1}{i(1+i)^n}$
PW	Present worth	The single value or worth today	—
PWF	Present worth factor	Converts a future amount to an amount today	$PWF = \dfrac{1}{(1+i)^n}$
SAF	Sinking fund annuity factor	Converts a future amount into an equivalent annuity	$SAF = \dfrac{1}{(1+i)^n - 1}$

*The two symbols have the following definitions:
n = the number of years in the evaluation period
i = the interest rate or other cost of money factor used

plete, long-term analysis is well worth considering in determining energy savings.

3.4.2 Marginal Cost Analysis. Marginal (or incremental) cost analysis is more a concept than an economic model. The marginal concept has predominant use in the economic community and has popular use in making decisions. Marginal cost analysis is simply the determination and use of the next increment of the cost of money or cost of electric energy. It is usually prudent to consider costs of the next increment of power, production, investment, and money.

3.4.3 Life Cycle Costing. Life cycle costing (LCC) is the evaluation of a proposal over a reasonable time period considering all pertinent costs and the time value of money. The evaluation can take the form of present value analysis, which this section will use, or uniform annual cost analysis.

The LCC method takes all costs and investments at their appropriate points in time and converts them to current costs. Inflation is assumed equal for all cost factors unless it is known to differ among cost items. Items for consideration include:

(1) Design cost
(2) Initial investment
(3) Overheads
(4) Annual maintenance costs
(5) Annual operating costs
(6) Recurring costs
(7) Energy costs
(8) Salvage values
(9) Economic life
(10) Tax credits

3.4.4 Example of Energy Economics. An understanding of the aforementioned concepts is best clarified by a moderately complex example. The example shown here is for illustration only and has no real life counterpart.

Mr. Smith shall decide either to keep his present car (A) or purchase one of two new cars (B or C). Car B costs $8000 and averages 36 m/g of gas. Car C costs $7000 and averages 30 m/g. His present car is worth $4000 (salvage) on a trade-in. He does not anticipate keeping his present car for more than three years. The maintenance costs on all three cars are shown below.

Maintenance	Car A	Car B	Car C
interval:	10 000 m	15 000 m	30 000 m
cost:	$200	$75	$100

Mr. Smith will have to install new equipment on his car this year at a cost of $200, and he will probably need additional work next year which currently costs $250. He anticipates a $100 expenditure on car B at 60 000 mi. It is anticipated that inflation will continue at a 15% yearly rate, but gasoline will increase 25% per year from a $1.30 per gallon level. Mr. Smith's cost of money is 18%.

Mr. Smith will have 120 000 mi on his present car in the third year, so this will be a sufficiently long period for evaluation.

Gasoline costs are calculated using the relationship of

miles/year · (gallons/mile) · (cost/gallon) = cost/year.

Solution:

$$\text{Fuel cost car A} = 20\ 000 \cdot \frac{1}{12} \cdot \$1.3$$
$$= \$2167 \text{ for year } 1$$

$$\text{Fuel cost car B} = 20\ 000 \cdot \frac{1}{36} \cdot \$1.3$$
$$= \$722 \text{ for year } 1$$

$$\text{Fuel cost car C} = 20\ 000 \cdot \frac{1}{30} \cdot \$1.3$$
$$= \$867 \text{ for year } 1$$

Table 5
Annual Cost Dispersions

Year	Item	Car A	Car B	Car C
1	Fuel	$2167	$ 722	$ 867
	Maint	200	75	0
	Major	200	0	0
2	Fuel	$2167 · 1.25 = $2709	$ 722 · 1.25 = $903	$ 867 · 1.25 = $1084
	Maint	$ 200 · 1.15 = $230	$ 75 · 1.15 = $86	$ 100 · 1.15 = $115
	Major	$ 250 · 1.15 = $287	$ 100 · 1.15 = $115	0
3	Fuel	$2167 · 1.5625 = $3386	$ 722 · 1.5625 = $1128	$ 867 · 1.5625 = $1355
	Maint	$ 200 · 1.3225 = $264	$ 150 · 1.3225 = $198	$ 100 · 1.3225 = $132
	Major	0	0	0
	Salvage	−1000	−3500	−2500

Table 6
Present Values of Annual Costs

Year	Car A	Car B	Car C
1	2567 · 0.847 = 2175	797 · 0.847 = 675	867 · 0.847 = 735
2	3226 · 0.718 = 2317	1104 · 0.718 = 793	1199 · 0.718 = 861
3	2650 · 0.609 = 1614	−2174 · 0.609 = −1324	−1013 · 0.609 = −617
Subtotal net future costs	$6106	$ 144	$ 979
Purchase (with 6% tax)	0	8480	7420
Trade-in	0	−4000	−4000
Total PW	$6106	$4624	$4399
Total expenditures w/o PW	$8443	$4207	$4473

Car A will have maintenance performed every 10 000 mi or twice each year at $100 each for $200 per year. Car B will have maintenance performed at 15 000, 30 000, 45 000, and 60 000 mi or in the first, second, and twice in the third year. Car C will have maintenance performed once in the second and third years due to the 30 000 mi maintenance interval.

Salvage at the end of the three years will be $1000 for A, $3500 for B, and $2500 for C.

Tables 5 and 6 were developed using the costs noted above. In Table 6, it can be seen that failure to use the life cycle costing method would have resulted in choice of Car B because of its higher trade-in value. Furthermore, keeping the present car would appear to be twice the cost of A instead of just 30% more. In any case, energy efficiency made the difference.

The car example covers the major components in an industrial project except overheads, design, and installa-

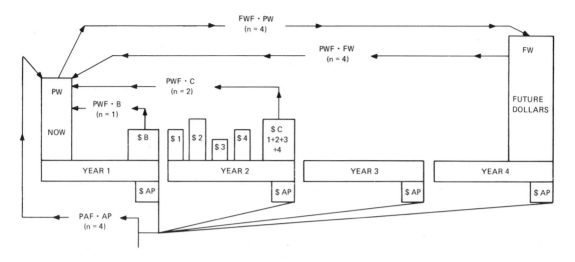

Fig 6
Time Value Chart

tion costs. The two tables used in the car example typify those required in an energy study. All pertinent costs should be listed by year and by alternative. Each cost is then appropriately inflated either in the original table or a second table. Finally, the costs are present valued in tabular form and totalled for evaluation. Sensitivity studies such as varying fuel cost should be done on separate tables to reduce the possibility of error.

3.5 Utility Rate Structures

3.5.1 An Electric Tariff. A tariff is filed by each electric company and approved in its filed form or as modified after rate hearings by the regulatory body. Each tariff has two sections:

(1) Rules and regulations
(2) Rate schedules

Rules and regulations are conditions under which a utility will supply power to a customer. This includes billing practices, rights-of-way, metering, continuity

of service, power factor, line extensions, temporary service, and many other details.

Rate schedules are the prices for electric service to different classes of customers. Four common classes are residential, commercial, industrial, and street or area lighting. There may be several rate schedules available on each customer class which are usually based on load magnitude. Special rate schedules and individual contracts are also common. A typical rate schedule for commerical and industrial customers usually contains most of the following elements: rate availability and characteristics; net rates for demand, energy and power factor; minimum charges; payment terms; terms of the contract; off-peak service; untransformed service and riders. Each electric rate is usually contracted for a period of one year and customers are entitled to the cheapest available rate, providing they meet the service characteristics specified in that rate. The following paragraphs describe the elements in a rate schedule and subsequent sections will cover detailed examples.

3.5.2 Rate Structure Elements. The usual rate structure establishes monthly charges for kilowatt demand, energy (kWh), and the power factor, which are added to comprise a base rate. Rate structures also include a fuel or energy adjustment charge which is applied to all kilowatthours consumed and then added to the base rate to obtain the total charge. The load factor, described more fully in Section 4, also affects the utility bill.

The demand component is designed to allow the utility to recover the capital costs associated with the construction of generating stations, substations, and transmission and distribution lines capable of meeting the customers' demand requirements. In most cases, a customer pays for the average demand in the highest 15 min or 30 min energy usage period during each billing period. Since there are 2880 fifteen minute periods in a month, one can easily see the reason for controlling demand. The utility shall supply sufficient capacity to meet this one period out of the total 2880 periods each month. Hence, both utilities and customers benefit from good demand control.

Energy charges are much easier to understand since one pays for the number of kilowatthours used to do the work required. A customer normally pays for all kilowatthours used. The energy component of a bill primarily recovers fuel costs but it also recovers operation and maintenance costs such as expendable materials, salaries and wages, gasoline, and tools.

Most utilities charge for reactive power usage (kilovars) for at least the very large users. The reactive power supplied to motors and transformers shall be paid for in some manner since the power company shall size its facilities to generate and transmit these kilovars. The total requirement is determined by the vector sum of the real and reactive power, so the term power-factor clause is commonly used. Methods used to calculate the reactive charge vary from a very clear charge per kilovar hour or per kilovar demand to what appear to be hidden means. The reactive charge can be reduced or eliminated by installing power-factor correction equipment — normally static capacitors.

The load factor is the relationship between the average kilowatt demand and the peak kilowatt demand. Utilities prefer a constant, nonvarying load or a 100% load factor where the average usage and the peak usage are the same. Many tariffs are structured to encourage better load factors.

The last common element of a rate structure is a fuel or energy adjustment clause (EAC). The purpose of this billing procedure is to enable a utility to recover its fuel costs quickly in a market where fuel costs fluctuate widely within short periods of time. The main objective of the EAC is to eliminate frequent and costly rate cases, an expense that is borne by each customer. The charge is normally applied to all kilowatthours used.

The following is an explanation of the rate forms most often used today. The rate structure atmosphere is changing so quickly that the following information can be outdated in a short period of time. Furthermore, the regulatory climate in each state is so different that these rate forms may no longer be available in some states.

(1) *Declining Block Rate.* The rate for the first kilowatt and kilowatthour is

typically the highest cost per kilowatt or kilowatthour. Hence, an initial kilowatt or kilowatthour block is billed at the highest rate. Additional consumption beyond this first block is then billed at a lower rate. There is no limit to the number of possible blocks, but it is unusual to see more than four.

This rate form was developed because utilities found that as a customer's consumption increased, the relative cost to provide the electric service decreased. The reduction in service costs was then reflected in a lower charge per unit as usage increased.

(2) *Demand Rates.* The determination of demand was covered in Paragraph 2 of this section. However, this rate sometimes includes the effects of load factor and is called a load factor, hours use, or demand-energy rate. In this rate form, the number of kilowatts in the first energy block is determined by kilowatt demand and a predetermined number of hours use, as shown in the example in 3.6. The number of kilowatthours in subsequent energy blocks is determined in the same manner. Thus, the larger the ratio between the average kilowatt and peak kilowatt, the more kilowatthours are billed in the lowest energy block and the lower a power bill becomes.

$$\text{Load factor} = \frac{\text{average kW (or kVA)}}{\text{peak kW (or kVA)}}$$

$$= \frac{kW_A}{kW_P}$$

$$\text{Hours use} = \frac{kWh}{\text{peak kW}}$$

$$= \frac{kW_A}{kW_P} \cdot \text{hours per month}$$

(3) *Seasonal Rates.* Power company yearly load patterns vary from one company to another. Some utilities experience a summer peak, while others see a winter peak. Other companies have yearly peaks that have no seasonal correlation. To discourage wasteful use of electricity in the peak seasons, some utility commissions require a higher charge during the peak seasons. In some rate schedules, the highest kilowatt demand during the peak season, or any month, determines the minimum kilowatt billing demand for the next 11 months or the next off-season months. This method of seasonal billing, sometimes called a ratchet clause, should encourage a large customer to use demand control or load management techniques during the peak season. The concept is based on the fact that the cost to provide service during the peak season is greater than at other times of the year.

(4) *Interruptible Rates.* While some individuals use interrruptible and curtailable interchangeability, this recommended practice will use interruptible to mean a rate based on the premise that the utility turns off the electric supply to a facility.

As a rule, interruptible rates are considerably lower than general service rates and, hence, have definite economic advantages. Sometimes the number of interruptible hours per year, or the number of hours per interruption, or both, are limited by the rate schedule. The customer shall weigh the benefit of the greatly reduced electrical costs against the losses associated with a complete shutdown.

(5) *Curtailable Rates.* In a curtailable rate structure, the customer makes predetermined, voluntary load reductions upon request by the utility. This rate structure usually involves some formal agreement between the user and the utility. The agreement usually involves such

important criteria as:

(1) The time period between the power company request for a load reduction and the reduction

(2) The magnitude of the reduction

(3) The maximum number of curtailments per year

(4) The maximum length of each curtailment

(5) The total number of curtailable hours in a year

This type of rate has advantages for the utility in that it can shed load quickly when critical power shortages occur. It should be noted that the customer may not have to completely shutdown to obtain a rate reduction. Generally, an interruptable rate will be lower than a curtailable rate. However, the rate may include an extremely heavy penalty charge for failure to curtail on request.

3.5.3 Proposed Electric Rate Structures. Conservationists, environmentalists, politicians, social scientists, and others looking for ways to reduce the growth in electricity usage and residential rates have influenced public utility commissions. Many factions believe that existing rate structures discriminate against certain customer classes and are a hindrance to their cause. Specifically, they believe the declining rate blocks and the lower unit cost for electricity enjoyed by commercial and industrial customers cause waste and unfairly discriminate against small users of electricity. Their natural solution is to change the rate structures. The purpose of this section is to familiarize the plant engineer with the more frequently proposed new rate structures.

Because new rate forms are often not cost based, companies, consultants, and engineers should participate in the rate-making process by intervening in rate cases. Utility commissions and the politi-

cal system should have the technical input of the engineering professions in the decision-making process.

The five fairly common rates proposed in 1982 were:

(1) *Flat Rates.* In this rate, all users of electricity would pay the same amount per kilowatthour for all kilowatthours used. In some cases, demand charges would also be eliminated.

(2) *Inverted Rates.* This rate form is the reverse of declining block rates. The unit cost would rise with higher usage. The first kilowatthours consumed would cost less per unit than the last kilowatthours used.

(3) *Marginal Rates (Marginal Cost Pricing) (MCP).* In this system, the utility would charge each customer based on the actual added cost that his usage imposes on the utility. The power company would first calculate how much each customer adds to its operating costs, then it would anticipate how its system shall be expanded to meet growing demand and assess all customers in proportion to their contribution to that expanding demand. A variation of MCP is what rate experts refer to as long-range incremental costing (LRIC). Pricing is based on the expected cost to produce electricity at some point in the future.

(4) *Lifeline Rates.* The basic concept of lifeline rates is to lower bills of low uers of electricity who are assumed to have low income. This is accomplished by inverting the rate structure for residential customers only. The revenue shortfalls are made up by higher rates for other customers.

(5) *Time-of-Day Rates.* This rate form would establish a pricing mechanism where electricity, kilowatts and kilowatthours, used during peak hours of the day would cost more than power used during the off-peak hours. The number of hours

in each period is determined by each power company. The main purpose of this rate is to smooth out the utility's load profile to improve the load factor. A load factor rate (demand-energy) is a form of time-of-day rates. Many power companies are currently experimenting with time-of-day rates.

The Public Utility Regulatory Policies Act of 1978 (PURPA) created new procedures for rate making by establishing Federal *standards* for rate redesign and by setting up new classes of intervenors. PURPA suggests that rate reform should encourage one or more of the following in addition to encouraging cogeneration[11] : conservation of energy, efficient use of utility facilities and resources, and equitable rates for electric customers. The effect of PURPA should be to encourage utility rate intervention by individuals. This action will probably cause additional proposals for new rate structures, and rate cases will be more drawn out.

3.6 Calculating the Cost of Electricity. A few examples will simplify the seemingly complex nature and wide variety of electric rates. In the 1980s one can expect wide use of block rates, demand usage rates, and fuel charges. Virtually all large utilities include a provision for reactive (var) charges in selected rates by using power factors or some form of demand or block rate, or both. The flat rate is not covered due to its simplicity. The utility or public utility commission will provide rate details for a specific situation.

3.6.1 Block Rate with var Charge Example. Tables 7 and 8 show rate schedules as they might be received from

[11] PURPA encourages cogeneration by setting general guidelines for sale of electricity back to utilities and for "back-up" service to cogenerators.

the utility commissions. The rate schedule is essentially a contract with a utility and should be read and understood. The riders and general rules and regulations are also part of the contract. For this example, assume that the plant's July electrical consumption is 2520 kW, 1 207 200 kWh, and 896 kvar which produces a power factor of 94.2%. Furthermore, the plant is billed on Schedule A (Table 7), and the fuel charge is 1.5¢/kWh (or 15 mil).

There are two blocks (Tables 7 and 8) for the kilowatt (kW) charge and a flat rate for the kilovar (reactive demand) charge. Notice the higher charge for summer usage which indicates that this is a summer peak utility. The first 50 kW is billed at $4.83/kW and the remaining 2470 kW at $3.80 for a total charge of $9627.50. It is important to note that the demand is determined for the current month only. The minimum charge on this rate is $11.00 plus fuel. The entire 896 kvar usage is billed at $0.20/kvar for a total reactive charge of $179.20.

All three blocks for kilowatthour charge are used for this load. The large portion of the kilowatthours in the last block is an indication that a different rate for higher usage may be available. The first 40 000 kWh are billed at 2.654¢/kWh, the next 60 000 kWh at 2.094¢, and the remainder at 1.524¢. The total kilowatthour charge is then $19 191.73.

The flat rate fuel charge of 1.5¢/kWh is applied to the entire 1 207 000 kWh usage. The fuel charge of $18 108 is then added to the demand and energy charges for a total bill of $47 106.43. The average cost of electricity is 3.9¢/kWh.

3.6.2 Demand Usage Rates Example. In the preceding section, the large number of kilowatthours in the last block

Table 7
Schedule A

Applicable to any commercial or industrial consumer having a demand equal to or in excess of 30 kW during the current month or any of the preceding 11 months.

Monthly Rates:
(1) Kilowatt Demand Charge

		Summer	Winter
		\$ per kW	
For the first	50 kW	4.83	4.01
For all excess over	50 kW	3.80	2.98

(2) Reactive Demand Charge

		¢ per kvar	
For each kvar of billing demand		20.0	20.0

(3) Kilowatthour Charge

		¢ per kWh	
For the first	40 000 kWh	2.654	2.354
For the next	60 000 kWh	2.094	1.794
For all excess		1.524	1.274

(4) Seasonal Rates: The winter rates specified above shall be applicable in seven consecutive monthly billing periods beginning with the November bills each year. The summer rates shall apply in all other billing periods.

(5) Fuel Cost Adjustment: The above kilowatthour charges shall be adjusted in accordance with the Fossil Fuel Cost Adjustment, Rider No 6.

(6) Other Applicable Riders: The rates specified above shall be modified in accordance with the provisions of the following applicable Riders:

Primary Metering Discount	Rider No 2
Supply Voltage Discount	Rider No 3
Direct Current Service	Rider No 5

Minimum Charge: \$11.00 per month or fraction of a month plus fuel cost adjustment.

Maximum Charge: If a consumer's use in any month is at such low load factor that the sum of the kilowatt demand, reactive demand and kilowatthour charges produces a rate in excess of 11.0¢ per kWh, the bill shall be reduced to that rate per kilowatthour of use in that month plus the fuel cost adjustment charge but not less than the minimum charge.

Special Rules:
(1) Combined Billing: Where two or more separate installations of different classes of service on the same premises are supplied separately with service connections within 10 ft of each other, the meter registrations shall be combined for billing purposes, unless the consumer shall make a written request for separate billing.

(2) Schedule Transfers: If for a period of twelve consecutive months, the demand of one installation or the undiversified total demand of several installations eligible for combined billing in each such month is less than 30 kW, subsequent service and billing shall be under the terms of the general commercial schedule for the duration that such scheduling is applicable.

(3) Reactive Billing Demand
(a) If the kilowatt demand on any class of service is less than 65 kW for three-phase installations or 75 kW for single-phase installations, the reactive billing demand shall be zero.
(b) If the kilowatt demand is 65 kW or higher for three-phase installations or 75 kW or higher for single-phase installations, the reactive billing demand shall be determined by multiplying the monthly kilowatt demand by the ratio of the monthly lagging reactive kilovoltampere hours to the monthly kilowatthours.

(4) Service Interrruption: Upon written notice and proof within ten days of any service interruption continuing longer than twenty-four hours, the company will make a pro rata reduction in the kilowatt demand rate. Otherwise the company will not be responsible for service interruptions.

Table 8
Schedule B

Applicable to any consumer having a demand of less than 10 000 kW and using more than 500 000 kWh per month during the current month or any of the preceding 11 months. No resale or redistribution of electricity to other users will be permitted under this schedule.

Monthly Rates

(1) Kilowatt Demand Charge

		Summer	Winter
		$ per kW	
For the first	50 kW	4.83	4.01
For all excess over	50 kW	3.80	2.98

(2) Reactive Demand Charge

	¢ per kvar	
For each kvar of billing demand	20.0	20.0

(3) Kilowatthour Charge

		¢ per kWh	
For the first	40 000 kWh	2.654	2.354
For the next	60 000 kWh	2.094	1.794
For the next	200 kWh per kWd but not less than 400 000 kWh	1.524	2.274
For the next	200 kWh per kWd	1.144	0.944
For all excess		1.004	0.794

(4) Seasonal Rates: The winter rates specified above shall be applicable in seven consecutive monthly billing periods beginning with the November bills each year. The summer rates shall apply in all other billing periods.

(5) Fuel Cost Adjustment: The above kilowatthour charges shall be adjusted in accordance with the fossil fuel cost adjustment Rider No 6.

(6) Other Applicable Riders: The rates specified above shall be modified in accordance with the provisions of the following applicable Riders:

Primary Metering Discount	Rider No 2
Supply Voltage Discount	Rider No 3
Consumer's Substation Discount	Rider No 4

Special Rules:

(1) Submetering or Redistribution Prohibited: This schedule is applicable only where all of the electricity supplied is used solely by the consumer for his own individual use.

(2) Schedule Transfers:

(a) If for a period of 12 consecutive months, the kilowatthour use in each such month is less than 500 000 kWh, subsequent service and billing shall be under the terms of Schedule A when this schedule is applicable.

(b) If in any month the maximum 30 min kW demand exceeds 10 000 kW, the consumer shall contract for service under Schedule B beginning with the next succeeding month.

(3) Reactive Billing Demand: The reactive billing demand shall be determined by multiplying the monthly kilowatt demand by the ratio of the monthly lagging reactive kilovoltampere hours to the monthly kilowatthours.

(4) Service Interruption: Upon written notice and proof within ten days of any service interruption continuing longer than twenty-four hours, the company will make a pro rata reduction in the kilowatt demand rate. Otherwise, the company will not be responsible for service interruptions.

Table 9
Riders

Rider No 1 — Fuel Adjustment for Special Contracts: The cost of fuel as used in Rider No 1 to Tariff PUCO No 11 shall be the delivered cost of fuel as recorded in Account Nos 501 and 547. Such fuel cost will be reported to the commission on a routine basis. Any proposed change in the type of fuel to be purchased, source of supply, or means of transportation which is estimated to increase or decrease the cost of fuel per million Btu by $0.01 or more shall be submitted to the commission for approval. Unless the commission shall take positive action within 15 working days to disapprove a proposal of an applicant, such proposal shall be deemed to have been approved.

Rider No 2 — Primary Metering Discount: If the electricity is metered on the primary side of the transformer, a discount of 2% of the primary meter registration in each of the company's electric schedules in which this rider is applicable will be allowed for electricity so metered.

Rider No 3 — Supply Voltage Discount: A discount on the monthly kilowatt demand charges in each of the company's electric schedules in which this rider is applicable will be allowed when the supply is entirely from 132 kV overhead circuits or 33 kV overhead circuits (for the purpose of this rider 33 kV overhead shall include 13.8 kV overhead transmission circuits fed directly from a power plant bus):

Class of Supply	Discount per kW of Demand Billed per Month
132 kV overhead	$0.30
33 kV overhead	$0.10

Rider No 4 — Consumer's Substation Discount: If the consumer elects to furnish and maintain or lease, or otherwise contract for all transforming, switching, and other equipment required on the consumer's premises, a discount of $0.30/kW or demand billed will be allowed on the monthly kilowatt demand charges in each of the company's electric schedules in which this rider is applicable.

Table 10
Block Rate Example

Usage: 2520 kW; 896 kvar
    1207 200 kWh

Rate: Schedule A
    Table 8

(1) Kilowatt demand charge
    50 kW · $4.83 =   251.50
    2470 kW · $3.80 = 9386.00
                  $9627.50

(2) Reactive demand charge
    896 · $0.20 = $179.20

(3) Kilowatthour charge
    40 000 · $0.02654 =   1061.60
    60 000 · $0.02094 =   1256.40
    1 107 200 · $0.01524 = 16 873.73

    Total charge: $19 191.73

(4) Fuel charge
    1 207 200 · $0.015 = $18 108

(5) Total electric charge
    kW =   9627.50
    kvar =   179.20
    kWh = 19 191.73
    Fuel = 18 108.00
    Total $47 106.43

indicated the possibility of a better rate. The usage is more applicable to the rate schedule noted in Tables 8 and 9; therefore, we will use the same usage figures on Schedule B. This rate schedule combines the block rate and demand usage rate. In addition, the customer can benefit from owning the equipment on his property (see Table 9, Rider no 4). The cost benefit of owning equipment to supply the 2500 kW (2640 kVA) plus load is $9000 per year.

The demand charges will still total $9806.70. The first two kilowatthour blocks will be the same at $1061.60 and $1256.40. The remaining 1 107 200 kWh will be allocated to the remaining blocks. It should be noted that a minimum of 400 000 kWh ($6096) is billed in the third block. The total number of kilowatthours in each demand usage block for the example is easily determined by multiplying the 200 kWh/kWd by the demand: 200 kWh/kWd · 2520 = 504 000.

To determine the hours in each block, the following table is used:

Total kilowatthours	1 207 200	40 000 in Block No 1
	− 40 000	
Balance for Block No 2	1 167 200	60 000 in Block No 2
	− 60 000	
Balance for Block No 3	1 107 200	504 000 in Block No 3
	− 504 000	
Balance for Block No 4	603 200	504 000 in Block No 4
	− 504 000	
Balance for Block No 5	99 200 kWh	99 200 in Block No 5

Table 11
Demand Rate Example

Usage: 2520 kW
        1 207 200 kWh
        896 kvar

Rate: Schedule B
        (Table 8)

(1) Kilowatt demand charge
        50 · $4.83 =   241.50
        2470 · $3.80 = 9386.00
                       9627.50
        Credit for Rider No 4 = $756.00
        ($0.30 · 2520)

(2) Reactive demand charge
        896 · $0.20 = $179.20

(3) Kilowatthour charge
        Block No 1   40 000 · $0.02654 = $ 1 061.60
        Block No 2   60 000 · $0.02094 =   1 256.40
        Block No 3   504 000 · $0.01524 =   7 680.96
        Block No 4   504 000 · $0.01144 =   5 765.76
        Block No 5   99 200 · $0.01004 =   995.97
                     1 207 200   16 760.69

(4) Fuel charge @ $1.5¢/kWh
        1 207 200 · 0.015 = 18 108.00

(5) Total electric charge
        kW   9627.50
        kvar   179.20
        kWh   16 760.69
        Fuel   18 108.00
        Total   44 675.39
        Possible credit   −756.00
        Total W/credit   43 919.39

The kilowatthours calculated for each block are multiplied by the appropriate rate to determine the total kilowatthour charge. For this schedule, the charge is $16 760.69, which is $2431.04 less than the amount using the previous schedule. The fuel charge is the same. The average cost per electricity is then 3.64¢/kWh.

3.6.3 Important Observations on the Electric Bill. Energy conservation demands a close look at energy costs. It is obvious from the preceding example that energy is the biggest portion of this bill. This fact is not always discernable as some rates include some minimum fuel cost in the kilowatthour portion of the rate. In the example, energy represents 40% of the electric cost. The costs associated with getting the power to the user represent another 38% of the bill. The cost to provide the facilities capable of supplying the load (demand cost) represent only 22% of the cost. However, in other cases, demand may represent the major part of the bill.

Since in this case each kilowatthour is associated with 78% (40 + 38) of the electric cost, a 10% reduction of kilowatthours brings four times the benefit over an equal reduction in demand. A 10% reduction in peak demand will reduce the power bill by only 2% which can be offset by a 2.5% increase in kilowatthours. In this case, the plant engineer should proceed with caution in controlling demand and, preferably, look for ways to reduce kilowatthours — even during off-peak hours.

Many articles and technical papers stress power-factor correction. In this case, the power factor is exceptionally good, but it is beneficial to see the savings achievable by power-factor correction. With an 80% power factor, the reactive demand is 1512 kvar (0.6 · 2520) and cost $302.40. Power correction to the 94% (896 kVA) level is then *worth* $123.20 per month (302.40 − 179.20) or approximately $1500 per year.

Finally, a reduction in usage will not give a proportional reduction in the electric bill. Suppose this manufacturer achieves a 10% reduction in kilowatthours, kilovars, and kilowatts, his savings is only 5% as shown in Table 12 due to the removal of the least costly increments.

3.7 Loss Evaluation

3.7.1 Introduction. All electrical equipment has some loss; nothing is 100% efficient. These losses can vary with output levels and age, or they can remain constant. For example, conductor losses

**Table 12
Dollar Savings From
Energy Reduction**

Energy

1 207 200 · 0.1 = 120 720 kWh saving
    99 200 kWh Block No 4 savings = $ 995.97
    21 520 kWh Block No 3 savings =   246.19
                                  $1242.16 energy savings

.kvar "flat" 0.1 · 179.20 = $17.92
.kW 0.1 · 2520 = 252 kW savings
252 Block No 2 savings = 957.60
Total savings = $2217.58 or 4.96%

vary as the square of the load current while the magnetic losses of a transformer are relatively constant.

In virtually all loss evaluations, a load profile shall be established either by analytical or by empirical methods. Furthermore, the efficiency of the device under investigation shall be determined for each set of anticipated operating conditions. The efficiency at full load is meaningless for comparing two devices that will be operated at half load unless the losses are solely a function of load.

All losses can be classified into two types. No-load loss is the quantity of losses when the device is idling or in a standby mode. Load losses are the additional or total level of losses at each load increment. The *efficiency* of a device is usually given at the full-load condition, which is but one point in the efficiency spectrum for many devices. Even the single point efficiency can have different values, depending on the standard under which the device was tested. (See [1].)

3.7.2 No-Load (or single value) Loss Evaluation. The no-load losses are constant for transformers, motors, and adjustable speed drives. The losses do vary as a function of voltage, frequency, and temperature, but these variables are expected to remain fairly constant over the evaluation period. Energy costs on lighting systems or similar processes with constant losses can be evaluated by using the no-load loss technique and substituting the "on" values for no-load values.

The cost of no-load losses has two components: the demand cost D_N and the energy cost Q_N. The demand cost is merely the diversified kilowatts or kilovoltamperes times the tail demand rate, the cost of the last block used in the demand charge. The energy cost is the hours of *on time* times the energy charge

which is the tail rate *including* the fuel charge.

The diversity factor is the per unit amount of an individual load that contributes to the billed demand. The diversity factor varies from zero when the load does not contribute to the plant's billing demand to 1 when the entire load is added to the plant's billing demand.

In mathematical form, the no-load cost of losses is as follows:

$$D_N = \left(\frac{\text{diversity}}{\text{factor}}\right) \cdot \frac{\$}{\text{kW}} \cdot 12 \text{ months}$$
$$= \text{demand cost per year per kW of no-load loss} \tag{Eq 1}$$

$$Q_N = \left(\frac{\text{no load}}{\text{hours}}\right) \cdot \frac{\$}{\text{kWh}}$$
$$= \text{energy cost per kW of no-load loss} \tag{Eq 2}$$

or Eq 2 can be rewritten:

$$Q_N = \frac{\text{no-load hours}}{\text{day}} \cdot \frac{\text{days}}{\text{week}} \cdot \frac{\text{weeks}}{\text{year}} \cdot \frac{\$}{\text{kWh}} \tag{Eq 3}$$

or

$$Q_N = \frac{\text{no load hours}}{\text{week}} \cdot \frac{\text{weeks}}{\text{year}} \cdot \frac{\$}{\text{kWh}} \tag{Eq 4}$$

3.7.3 Load Loss Evaluation. Load losses are more complex only because they vary over the evaluation period; so they shall be developed in increments. No-load losses can be combined with load losses (for example, in cases where it is not possible to separate the losses or when it is desirable to look at total losses as a single entity). It is usually easier to treat load and no-load losses separately for motors and transformers because load losses can be mathematically expressed in terms of load.

Load losses have a demand and an energy component. The demand component once again includes the device's effect on the plant's electrical peak. The demand-loss cost for one year is then

$$D_L = \sum_{i=1}^{12} \left(\begin{array}{c}\text{diversity}\\ \text{factor}\end{array}\right)_i \cdot (\$/\text{kW})_i \cdot (P_i)$$

where
  i = month
  P_i = unitized level of load losses in the ith period

If the cost, relative power level and diversity factor are constant the equation is

$$D_L = 12 \cdot DF \cdot DC$$
  = demand cost per year per kW of load loss

where
  DF = diversity factor
  DC = demand cost

(Eq 5)

Since the energy cost is a function of the load cycling, the load cycle shall be determined. Meters can be installed on existing equipment to obtain actual values. If actual data is not available, a load schedule or profile can be obtained by metering similar process or by theoretical analysis. It is usually wise to group loads in terms of hours. This results in small loss of accuracy and great ease of calculation.

The energy cost is then the sum of energy used times the cost of energy. The sum of usage is simply the sum of the load levels times their associated on times. The number of hours per year is then multiplied by the tail energy rate (including the fuel charge). In some cases, the calculation may require summing several different types of load cycles during a year. In the simplest case of constant daily load, the following equation applies:

$$Q_L = 365 \cdot \frac{\$}{\text{kWh}} \cdot \sum_{i=1}^{n} t_i P_i$$

  = annual $/kW load loss

where
  n = number of periods
  t_i = duration of the ith period, hours
  P_i = unitized level of load losses in the ith period

(Eq 6)

In the next sections, the value of P_i is unitized in terms of load level and Q is put in terms of dollars per kilowatt per year. The loss evaluation generally is made for a period greater than one year but not exceeding the anticipated useful life of the equipment.

3.7.4 Motor Loss Evaluation with Example. The load and no-load losses in a motor combine in a manner shown in Fig 7. Without specific points on the curve, one can approximate motor losses by the no-load and load losses. The no-load losses are composed of the hysteresis, eddy current, and windage and friction losses at full-load speed and temperature. The no-load losses can, therefore, be caculated by using Eqs 1 and 2. The load losses (the remainder) varies as the square of the motor load.

The motor-load losses can be calculated by recognizing that the major component is the I^2R losses in the winding and the armature. Sophisticated programs would consider additional adjustments due to heating effects, etc. The value of load loss at any particular load is the square of the ratio of that load to nameplate. More specifically, the value of yearly motor load losses is as follows:

$$D_L = \sum_{i=1}^{12} \left(\begin{array}{c}\text{Diversity}\\ \text{Factor}\end{array}\right)\left(\frac{HP_i}{HPR}\right) \cdot DC_i$$

  = $/kW of annual load loss

$$Q_L = \sum_{i=1}^{n} t_i \left(\frac{HP_i}{HPR}\right)^2 \cdot EC_i$$

= \$/kW of load loss per period

(Eq 7)

where

D_L = demand cost per kW of load loss
Q_L = energy cost per kW of load loss
HP = peak motor load, hp
HP_i = motor load for the ith interval, hp
HPR = rated motor horsepower
n = number of intervals being evaluated
t_i = duration of the ith interval, hours
DC_i = \$/kW demand cost for the ith month
EC_i = \$/kWh energy cost for the ith interval

An example will illustrate the loss evaluation technique. Assume a motor is used in a five-day, ten-hour per day process that runs 50 weeks a year. The motor runs at 0.25 load for 2 h, 0.50 load for 4 h, full load for 2 h and idles the remaining 2 h. The tail rate energy cost is \$0.10/kWh and the tail demand rate is \$15.00/kW. The peak motor load is coincidental with the plant's electrical peak.

(1) The annual no-load energy cost is as follows:

Q_N = (10 h) · (5 days/week)
      · (50 weeks/year) · \$0.1/kWh
      = \$250/kW of no-load losses
D_N = 1 · 15 · 12 = \$180/kW

(2) The annual energy load loss is as follows:

Q_L = [(2 h) (0.25 load)^2
      + 4 h (0.50 load)^2 + 2h (full load)^2]
      · (5 days/week) · (50 weeks/year)
      · (\$0.1/kWh)

= (0.125 + 1 + 2) · 5 · 50 · 0.1
= 3.125 · 5 · 50 · 0.1
Q_L = \$78/kW of load loss

NOTE: No entry is required for the 14 h when the motor is off or for the 2 h when it idles.

(3) The annual demand load-loss cost is:

D_L = 1 · 15 · 12 = \$180/kW

A no-load and load motor loss reduction of 1 kW each is worth \$688 (250 + 180 + 78 + 180). With a five-year life and a 20% cost of money, each kilowatt reduction in both load and no-load losses is worth \$2058 more in purchase price for the aforementioned loading and electric rate.

3.7.5 Transformer Loss and Example. The transformer loss evaluation is almost identical to the motor evaluation. Important differences include the fact that losses are a squared function of kilovolt-ampere load, the no-load losses occur continuously, and the load usually increases each year. Additional sophistication can be added to show the effects of temperature due to load level and voltage level. The transformer can also be loaded well above nameplate in certain situations and it has longer life than a motor.

The equations are similar to the motor equations and the no-load demand equation is identical. The equations for transformer loss evaluation are as follows:

D_N = (see Eq 1)
Q_N = 24 · 365 · EC = 8760 EC

(Eq 8)

$$D_L = \sum_{i=1}^{n} \left(\begin{array}{c}\text{diversity}\\\text{factor}\end{array}\right) \left(\frac{kVA_i}{kVAN}\right)^2 \cdot DC_i$$

(Eq 9)

$$Q_L = EC \cdot \sum_{i=1}^{n} t_i \left(\frac{kVA_i}{kVAN}\right)^2$$

(Eq 10)

where

DC_i = demand cost in \$/kW for the ith month

EC_i = energy cost in \$/kWh for the ith month

D_N = no-load demand loss cost per kilowatt per year

Q_N = no-load energy loss cost per kilowatt per year

D_L = load demand cost per kilowatt per year

Q_L = load energy cost per kilowatt per year

kVAN = transformer nameplate rating, kVA

kVA$_i$ = kVA load for the ith interval

i = an interval of constant load

The effect of load growth can easily be added to the above equations by the following equation:

load-loss cost in year $n = (D_L + Q_L) \cdot [(1 + I)^{n-1}]^2$

(Eq 11)

Where I is the per unit rate of load expansion. This multiplication factor does not apply to the no-load cost of losses.

EXAMPLE: Consider the purchase of a 5000 kVA, 34 500–4160 V, 3ϕ transformer. The peak load is 3000 and should grow 5% per year. The tail-rate demand is \$10 and the tail-rate energy is \$0.10/kWh, and they are expected to increase 12% per year. Transformer peak and billing peak are coincidental. The factory operates in 2 shifts for 5 days each week all year. On evenings and weekends the load is approximately 40% of peak. During the 16 h of production, 4 h see 60% of peak, 4 h see 80% of peak, and the remaining 8 h are at 3000 kVA (note that kVA is used and not kW). Furthermore, the load curve is identical for all working days (which is usually not the case).

This transformer has a 30 year life at the specified loading, but if the load grows at 5% each year, this transformer will be replaced much sooner than 30 y if it is charged out at nameplate (5000 kVA). The years until the load reaches nameplate value can be calculated quite easily as follows:

$$\frac{5000}{3000} = (1 + 0.05)^n$$

Since

$$(1 + 0.05)^n = \frac{5000}{3000}$$

$$n \cdot \ln(1.05) = \ln\left(\frac{5}{3}\right)$$

$$n = \frac{\ln\left(\frac{5}{3}\right)}{\ln(1.05)} = \frac{0.51083}{0.04879}$$

= 10.46 or approximately 10 years

Therefore, the evaluation period should be 10 y if it is desirable to charge the unit out at nameplate loading.

(1) No-load losses for Year 1 (assuming equal load all year and constant rates):

$D_N = 1 \cdot 12 \cdot \$10 = (\$120/kW)/year$
$Q_N = 24 \cdot 365 \cdot 0.1 = (\$876/kW)year$

Total no-load loss for Year 1 is
120 + 876 = \$996/kW

(2) The load curve is as follows:

(a) Evenings and weekends are 0.4 · 3000 = 1200 kVA for 8 h per day (24 – 16 = 8) for 6 days and 24 h on Sunday for a daily equivalent of 48 + 24 = 72 h per week.

(b) The other levels are 0.6 · 3000 = 1800, 0.8 · 3000 = 2400 and 3000. At 6 · 4 = 24 and 6 · 8 = 48 h per week respectively.

(3) Load losses for Year 1:

$$D_L = 1 \cdot 12 \cdot \left(\frac{3000}{5000}\right)^2 \cdot \$10$$

$$= (\$43.20/\text{kW})/\text{year}$$

using weekly loads

$$Q_L = \left[72\left(\frac{1200}{5000}\right)^2 + 24\left(\frac{1800}{5000}\right)^2 \right.$$

$$\left. + 24\left(\frac{2400}{5000}\right)^2 + 48\left(\frac{3000}{5000}\right)^2 \right]$$

$$\cdot 52 \cdot \$0.10$$

$$= (4.1472 + 3.1104 + 5.5296$$
$$+17.28) \cdot 52 \cdot 0.1$$
$$= 30.0672 \cdot 52 \cdot 0.1$$
$$= (156.35/\text{kW})/\text{year}$$

The load losses for Year 1 cost $199.55.

(4) The losses in Year 9 are as follows:
No-load losses change only by inflation
or $996 \cdot (1.12)^9 = \$2762$
The load losses change by both inflation
and load or $(199.55) [(1.1)^8]^2 \cdot (1.12)^9 = \$199.55 (2.14)^2 \cdot (2.77) = \$2531$

3.7.6 Other Equipment. The method used in 3.7.3 and 3.7.4 can be used to evaluate any item or process cost. One needs only to determine the losses from no load to full load and the load cycle. Loss evaluation can apply to conductor sizing, rectification equipment, variable speed drives, lighting systems, controls and sources, and even different types of processes.

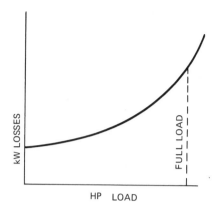

Fig 7
Motor Losses

3.8 Bibliography

[1] BONNETT, A. H. Understanding Efficiency in Squirrel-Cage Induction Motors, *IEEE Transactions on Industry Applications*, vol 1A-16, no 4, July/Aug 1980, pp 476-483.

[2] BROWN, Robert J. and YANUCK, Rudolph R. *Life Cycle Costing*, PA, Commonwealth of Pennsylvania, 1979.

[3] GRANT, E. L. and IRESON, W. G. *Principles of Engineering Economy*, 4th ed, New York, The Ronald Press Company, 1960.

[4] KOVACS, J. P. Economic Considerations of Power Transformer Selection and Operation, *IEEE Transactions on Industry Applications*, vol 1A-16, no 5, Sept/Oct 1980, pp 595-599.

4. Load Management

4.1 Introduction. Practicing good electrical energy conservation with existing plant equipment is considered load management. Many people narrowly define load management as demand control, but good engineering dictates control of electrical usage 24 hours each day and 365 days each year. This concept demands load curtailment even during the lowest usage times of the plant's load cycle.

The application of good load and demand control concepts requires an understanding of utility rates, auditing, and metering in addition to a basic knowledge of the process and load being controlled. The engineer shall audit and meter the system and then determine which of the electrical loads can be shed. With the above information, the load management program can begin.

4.2 Demand Control Techniques. Plant management may not necessarily be interested in trimming power peaks.

Their job is to optimize profits without sacrificing production quality or quantity. However, they can directly reduce the demand and energy charge without sacrificing product quality or quantity. The fundamental principle of demand control is fairly simple. It is necessary to determine at what time of the day and on which days the peak occurs and then determine which loads are in use at that time. Next the magnitude of the loads shall be determined. Decisions can then be made as to which operations can be curtailed or deferred to reduce the demand peak and the power bill. The subsequent sections show effective means of controlling plant demand.

4.3 Manual Methods. Maximum benefits can be attained from a plant power survey focused on those areas where it will pay to have continuous records made through the installation of permanent instrumentation. Total plant records will show how each piece of equipment contributes to the total load picture and will reveal whether equipment is operating within specifications. Areas of energy waste should be identified.

If certain heating, ventilating, and air-conditioning (HVAC) or other nonessential systems in the plant can be shut down when the building is unoccupied or for a few minutes during the peak demand period, a timer may be a very effective demand controller. Outdoor and indoor display lighting systems are also candidates for time-clock control. Electric process heat systems can also be timed or staged. All time switch based systems have limitations. Unless an astronomic or 7-day type is used, the controller shall be reset frequently. A power interruption can require subsequent time reset. Newer programmable timing controllers can provide hundreds of patterns and time changes during the year.

Interlocks prevent either one unit or several units from being operated with its pair of units or counterparts, simultaneously. The method has the advantage of being low in cost and fast to install. Heat treating operations are excellent candidates for interlock because of the high initial electrical usage to bring the unit up to temperature.

With proper metering, demand can be controlled manually by simply watching the meters. Ideally, local readings should be transmitted to a common point where one person (or even some device) observes the rate of consumption. This individual initiates load removal in a preplanned manner, that is to say, he drops the noncritical load to keep the demand under a predetermined level. A reliable communication system or even a remote control should be installed for the observer to initiate the turn-off and turn-on action. The observer needs to pay attention only during peak demand periods. An observer is not required at night for a one-shift, daytime operation. The observer will be able to predict usage patterns and peak times after on-the-job training. In essence, he will become a programmed predictive controller.

The advantages of manual control are that it shows what can be done in cutting demand, investment is small, and it enables management to think out the problems they will eventually have to face if they go on to some form of automatic load shedding.

4.4 Automatic Controllers. Although the previous methods and equipment might provide acceptable solutions for some applications, they have inherent disadvantages and do not lend themselves to complex, fine control. When a more complex, finer tuned operation is desired, a more sophisticated, automatic demand controller should be installed.

Automatic controllers can be categorized by operating principle: instantaneous demand, ideal rate, converging rate, predicted demand, and continuous integral. Some controllers are offshoots of these five basic versions; others are hybrids embracing more than one operating principle. Installation costs will vary depending on controller location and the number and location of controlled loads.

Most demand controllers require pulse signal inputs derived from the utility's demand meter. One pulse indicates the usage and the other pulse indicates the end of the demand interval. The controller then observes each interval for rate of usage. The following detailed descriptions will tell how each type of controller uses this information to control demand.

With an instantaneous demand controller [Fig 8(b)], action is taken when instantaneous demand exceeds the established prescribed setpoint value. A setpoint value is determined for the demand interval. Straight line accumulation or constant usage is presumed. When one-fourth of the demand interval has transpired, accumulated demand or kilowatthour usage should be no more than 25%. Loads are switched in and out of service in accordance with this criteria. This mode of operation might result in short cycling of loads. In any demand control system, short cycling can be effectively damped out by simply installing cycle timers in the control circuits of problem equipment or by having logic in the control that performs the same timing function.

With an ideal rate controller [Fig 8(c)], ultimate demand limit is prescribed, and a slope is established to define when usage indicates that this limit is likely to be exceeded. The ideal rate controller does not begin from *zero* at the start of the demand interval, but from an established offset point that takes into account nondiscretionary loads. Slope of the *ideal rate of use* curve is then defined by this offset point and a chosen maximum demand. The offset provides a buffer against unnecessary action early in the demand interval, thereby reducing cyclic equipment operation.

The converging rate controller [Fig 8(d)] works like the ideal rate controller, but it operates on an accumulated usage curve whose upper limit is defined by the specified maximum demand. It also employs an offset to minimize nuisance operation early in the demand interval. But unlike the ideal rate controller, which established parallel rate-of-use lines for loads, the converging rate controller load lines are not parallel.

They converge at the maximum demand point to permit vernier-like control toward the end of the demand interval when accumulated registered demand might be critically near the setpoint.

With a predicted demand controller [Fig 8(e)], average usage is observed periodically through the demand interval and compared with the instantaneous usage at that particular moment. This information is used to continually develop a curve of predicted usage for the remainder of the interval. If the projected curve indicates that the target setpoint might be exceeded, action is taken.

The continuous integral controller monitors power usage continuously, rather than only when a time pulse signal is transmitted by the utility company's demand meter. When action is called for, the controller activates a satellite cycle timer, which sheds loads for a predetermined, fixed period. If further action is called for, other timers are activated until usage is brought in line with the desired objective. Because satellite timers, once activated, will shed loads through a complete cycle and overlap the demand intervals, short-cycle operation is reduced.

4.5 Microcomputer System. By a keyboard input to a computer, the plant engineer can specify the maximum demand which can be tolerated, based on previous experience. The microcomputer continually monitors the plant's electric consumption and, by one of the aforementioned methods, determines if the demand limit will be exceeded. If demand is well below the limit, control action is not taken. If the computer predicts that demand may exceed the limit, the preselected loads can be automatically turned off to reduce demand or an alarm can alert the building engineer who then makes a manual adjustment decision. It provides several levels of load shedding priority. Some loads are designated *low priority*; these will be shed in round-robin rotation as needed. Others can be placed in a separate *high priority* category, which will not be shed until the supply of low-priority loads is exhausted. Limiting and load shedding can be increased during peak periods, with demand limits relaxed during low cost, off-peak times. Two of the most valuable features of this system are its logging/printing capability and the ability for in-house reprogramming.

With time-of-day control, equipment managed by the microcomputer system operates only when needed. This system can cycle the various building loads and can vary the load's duty cycles according to the time-of-day, staggering equipment off times to reduce electrical demand. Flexible load cycling thus retains full equipment capacity for fast warm-up or cool down, while reducing demand and energy expense under routine operating conditions.

4.6 Computerized Energy Management Control System. A central in-plant energy management system shall consider the total plant site, including fuel usage in the production of process steam and electricity, energy usage in the production process in maintaining the plant environment, and the cost of outside purchased electricity. This energy management system consists of three hierarchical tiers: the operating, supervisory, and management planning levels (Fig 8). The data within this hierarchy requires more manipulation and refining as it progresses from a lower to a higher level.

The uses for energy management systems are as diverse as the process they serve. The following subsections represent the more obvious ones:

4.6.1 Energy Distribution. By monitoring power production and purchases, engineers can recommend changes in the distribution or purchase of energy.

4.6.2 Monitor Energy Consumption. Constant monitoring and evaluation of energy usage by department or area can prevent extraordinary consumption.

4.6.3 Methods of Conservation. Computer analysis of plant conditions (cycling of exhaust fans, setting of air conditions, etc) can substantially minimize demand and, sometimes, consumption.

4.6.4 Load Shedding. Selective shedding of low-priority loads can minimize demand peaks. Load shedding is also valuable when utilities place absolute ceilings or penalties for excess use on the amount of energy supplied.

4.6.5 Cogeneration. A computer-based energy controller is beneficial for a plant using process steam to generate power for internal use. If by-product gas is to be burned as part of the fuel, the controller determines the economic mix of purchased oil with waste gas needed to produce the steam necessary for efficient power generation. The controller calculates that level of generation to be maintained under each operating condition and considers how the contract with the utility affects the mix.

4.6.6 Maintenance Prediction. Early detection of rising temperatures, abnormal currents, or other operating irregularities during normal monitoring by the computer can signal a need for maintenance before equipment is damaged. The savings in maintenance costs can be significant. With advance warning, equipment shutdowns can be scheduled to avoid costly disruptions of the production process.

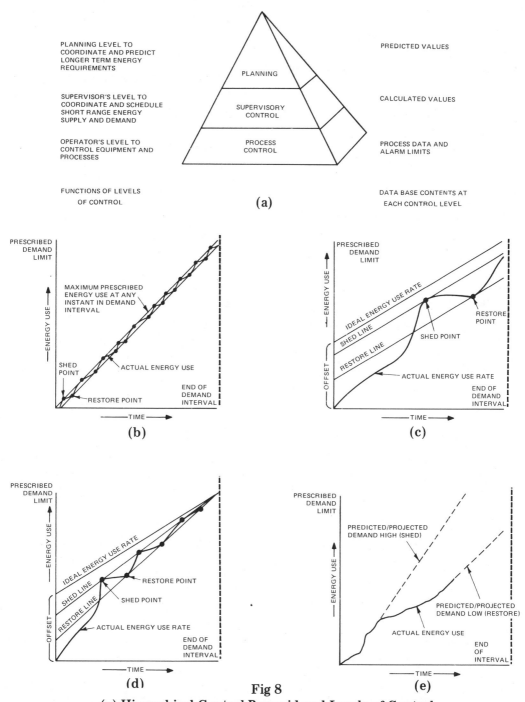

Fig 8
(a) Hierarchical Control Pyramid and Levels of Control
(b) Instantaneous Demand Principle   (c) Ideal Rate Principle
(d) Converging Rate Principle   (e) Predicted Demand Principle

4.7 Economic Justification for Energy Management Systems. An energy management system can provide substantial savings, either through reduced energy costs, or through increased production without corresponding increases in energy requirements. For complex systems, ROI (return on investment) from decreased costs and increased revenue may not alone be sufficient to justify the initial investment within the company's specified time. Many companies can justify energy management systems on an ROI basis, when the concept of lost production due to shortages or cutoffs is quantified and factored into the ROI analysis.

Today's management shall make investments to optimize the use of energy. Some industries may be required by the government to reduce energy demands. Other industries will have to prove that their use of energy already meets or exceeds the industry's standards for efficiency in terms of energy consumed per unit of product.

Additional benefits can be realized from a computer-based energy management system by allowing the computer to perform other duties. The data acquisition and reporting capability of the computer may be used to monitor and record the operation of pollution control equipment. Its ability to schedule preventive maintenance may be used to protect all major plant equipment, not just that connected with energy production or use. Furthermore, computers have a long history of handling accounting, payrolls, scheduling, etc. Energy management may prove economical where done in conjunction with these nonengineering functions.

4.8 Bibliography

[1] CHEN, Kao and PALKO, Ed. An Update on Rate Reform and Power Demand Control, *IEEE-IAS Transactions*, vol 1A-15, no 2, Mar/Apr 1979.

[2] DACQUISTO, J. F. Beating Those Power Demand Charges, *Plant Engineering:* Technical Publishing Co, Barrington, IL. Nov 1971.

[3] HANSEN, A. G. Microcomputer Building Control Systems Managing Electrical Demand and Energy, *Building Operating Management*, Trade Press Publishing Co, July 1977.

[4] HUGUS, F. R. Shipbuilding and Repair Facility Controls Demand to Reduce the Cost of Electricity, *Electrical Construction and Maintenance:* McGraw-Hill: New York, NY. July 1973.

[5] JARSULIC, N. P. and YORKSIE, D. S. Energy Management Control Systems, *Energy Management Seminar Proceedings*, Industry Applications Society, 77CH1276-51A, Oct 1976 and Oct 1977.

[6] QUINN, G. C. and KNISLEY, J. R. Controlling Electrical Demand, *Electrical Construction and Maintenance:* McGraw-Hill: New York, NY. June 1976.

[7] MAYNARD, T. E. Electric Utility Rate Analysis, *Plant Engineering:* Barrington, IL. Nov 1975.

[8] MECKLER, Milton, Energy Management by Objective, *Buildings*, Stamats Communications Inc, Cedar Rapids, Iowa. Nov 1977.

[9] NIEMANN, R. A. Controlled Electrical Demand, *Power Engineering:* Technical Publishing Co, Barrington, IL. Jan 1965.

[10] OCHS, H. T. Jr, Utility Rate Structures, *Power Engineering*, Barrington, IL. Jan 1968.

[11] PALKO, Ed. Saving Money through

Electric Power Demand Control, *Plant Engineering:* Barrington, IL. March 1975.

[12] PALKO, Ed. Preparing for the All-Electric Industry Economy, *Plant Engineering:* Barrington, IL. June 1976.

[13] PEACH, Norman. Do You Understand Demand Charges, *Power:* McGraw-Hill: New York, NY. Sept 1970.

[14] PEACH, Norman. Electrical Demand Can be Controlled, *Power:* New York, NY. Nov 1970.

[15] REKSTAD, G. M. Why You'll Be Paying More For Electricity, *Factory Management*, McGraw-Hill: New York. Feb 1977.

[16] RELICK, W. J. Using Graphic Instruments to Hold Down the Electric Power Bill, *Plant Engineering:* Barrington, IL. May 1974.

[17] WRIGHT, A. Keeping that Electric Power Bill Under Control, *Plant Engineering:* Barrington, IL. June 1974.

[18] Edison Electric Institute Information Service, October 26, 1976.

[19] S122 "Electric Utility Rate Reform and Regulatory Improvement Act," January 10, 1977.

[20] Computer Slashes Electric Bill, *Factory*, Sept 1974.

5. Conservation Considerations in Electrical Machines and Equipment

5.1 Induction Motors. There are three components to the electrical energy required by a motor:

(1) The mechanical load on the motor
(2) The mechanical losses in the motor
(3) The electrical losses in the motor and the electrical supply system

Item (3), electrical losses, is a function of the electrical environment, the nature of the load, and the design of the motor. Motor efficiency is a convenient way of relating these losses to the productive work being done by the motor. However, when a motor is not operating under constant load, efficiency requires redefinition to be meaningful. Consider a motor which is idling 80% of the time and loaded to 150% of nameplate 20% of the time. The rated efficiency of such a motor may have little relation to the net efficiency over the full cycle. Thus, a preferred definition for efficiency is:

$$\text{Efficiency} = \frac{\left(\begin{array}{c} \text{total} \\ \text{output*} \end{array}\right) \cdot (0.746/\text{cycle})}{\text{total input*/cycle}}$$

---

*Output is in horsepower hours and input is in kilowatthours.

The cycle may last only a few seconds, as on a punch press, or 15 min as on a material moving system, or perhaps for a full work shift where the load is constant except for rest periods and shift changeovers. In all cases, motor efficiency and, hence, operating costs should be evaluated over a full-load cycle when minimum energy costs are the objective.

The life of electrical insulation is a function of the operating temperature. A 10 °C increase in temperature cuts the insulation life in half and, correspondingly, a reduction of 10 °C doubles the life. In many instances, life is not a problem and, hence, little thought is given to the temperature of the equipment. However, temperature also affects the resistance of the windings of all electrical equipment. Thus, a cooler piece of equipment will create fewer losses. For example, the resistance of a conductor changes with temperature

$$Rt_2 = Rt_1 \frac{(M + t_2)}{(M + t_1)}$$

where

Rt_1, Rt_2, t_1 and t_2 = respective dc resistances and temperatures in °C, and M is a constant: 241 for copper and 228 for aluminum. Thus, a 10 °C reduction in motor temperature will reduce the dc resistance losses of the conductors by 3%–4%. A similar, though more complex relation, exists for the magnetic losses.

The efficiency of all electrical equipment has some sensitivity to the supply voltage magnitude, phase balance, wave shape, and frequency. Motors and transformers are designed to meet the specified temperature rise within a voltage range of ±10% of nameplate. This requires adequate-sized conductors to carry the current at −10% voltage and an adequate magnetic circuit for the +10% voltage level. As a consequence of this design requirement, a motor will operate slightly cooler and with lower losses than stated at a voltage level a few percentage points above the rated value at full load.

When variations in frequency shall be considered, as on a small isolated system, efficiency is more directly related to volts per hertz than to simple voltage. A constant volts per hertz ratio results in constant flux density and this is the significant parameter, not voltage.

Voltage phase unbalance, flicker, and wave distortions increase system and equipment losses more than generally recognized. See [6][12]. These losses are the result of two effects:

(1) Current flowing in conducting components not designed to carry current continuously. This is most likely to occur in synchronous machines, in wye-wye connected transformers, and in core-type transformers.

(2) Skin effects which occur mostly

in large conductors. The effective resistance to harmonic currents, which exist routinely in modern industrial plants, may be 150%–200% or more of the 60 Hz resistance. This effect is such that even an rms ammeter will not give a true measure of the heating effect of distorted currents flowing in large conductors.

By far the largest skin effect occurs in the rotors of motors. These rotors are designed for dc currents as in synchronous motors or the 1 Hz to 2 Hz slip frequency in induction motors. Any voltage imbalance results in a high-frequency current of twice line frequency (less the slip) flowing in the rotor. Small voltage imbalance can cause large currents to flow because they are only impeded by the negative sequence impedance of the motor. This impedance value is very near the locked rotor impedance of the motor. Thus, a 5% voltage imbalance may cause a 20%–30% current imbalance. The effective resistance of the rotor to this 120 Hz frequency will be approximately 5–8 times the dc resistance. This is caused by the skin effect of these large rotor bars accentuated by the closeness of the rotor magnetic structure. A 5% voltage imbalance can thus lead to a 50% increase in motor full-load losses. This phenomenon is not related to the load on the machine, but is a constant number of watts for a given voltage imbalance. Thus, the losses on an idling machine may be greater than assumed by a factor of two or more if there is significant voltage imbalance or harmonic distortion.

The ratings and prescribed characteristics of electrical equipment as specified in various IEEE, ANSI, and NEMA standards are not based on maximizing the utilization of electrical energy. While it is generally poor policy to exceed the stan-

---

[12] The numbers in brackets correspond to those of the Bibliography, 5.12.

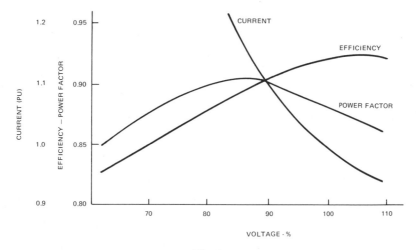

Fig 9
Effect of Voltage on Motors at Full Load

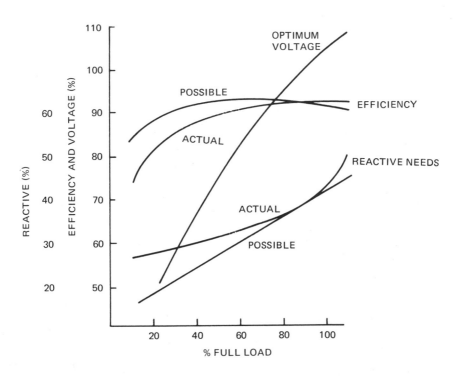

Fig 10
Motor Characteristics at Various Loads

dard ratings, it is not so obvious what the consequences are of underutilizing the various types of equipment. As noted above, efficiencies should be calculated over a full operating cycle when determining the actual utilization of energy. Other environmental considerations are given in this section for various pieces of equipment.

Induction motor efficiency is closely related to terminal voltage. Figure 9 shows this for a fully loaded typical motor with required current and resulting power factor. Note that efficiency peaks at over 100% voltage, but that maximum power factor occurs at approximately 85% voltage. Note also that minimum current does not occur at maximum power factor as is frequently assumed.

Figure 10 shows the benefit in regulating the motor voltage to the optimum value for the particular load on the machine. Note the possibility for improving the efficiency at light load by lowering the machine voltage. This voltage reduction technique is used in an energy-saving device, but three cautions shall be considered.

(1) If the machine is subject to a suddenly applied load, it may stall if the voltage is not first increased to provide the needed torque.

(2) If the voltage is regulated to the optimum value with a thyristor or other device that creates wave distortion, the increase in losses due to the distortion may exceed the reduction in the 60 Hz losses.

(3) When capacitors are used to improve power factor, the voltage will also be increased. This will improve the efficiency of a fully loaded motor, but it will reduce the efficiency of a partially loaded motor according to Figs 9 and 10.

The mechanism whereby voltage im-

Table 13
Effect of Voltage Imbalance on a
200 hp Motor at Full Load

Voltage imbalance (%)	0	2.0	3.5	5.0
Increase in losses (%)	0	8	25	50
Temperature rise (°C)				
Class A	60	65	75	90
Class B	80	86	100	120

balance causes an increase in motor losses was given above. Quantitatively, this effect is given in Table 13 for a typical (200 hp) motor.

Harmonic distortions will tend to increase losses comparable to those of Table 13. While the motor impedance increases at the distortion frequency and thus reduces the current flow for a given applied voltage, the skin effect is greater at these higher frequencies as are the losses in the magnetic materials. Thus, the approximate relationship of Table 13 is an adequate rule of thumb for harmonic voltages and simple phase imbalance.

Voltage flicker as produced by intermittent loads, welders, and electric arc furnaces causes additional losses in motors. This occurs for two reasons.

(1) The individual flickers are single-phase phenomena. Thus, they constitute a voltage imbalance with resulting losses as described. A 10% flicker occurring 50% of the time (on one phase or another) will have the effect of a constant imbalance of approximately 5% with a resulting additional motor loss of approximately 5% of motor rating.

(2) When the flicker is three phase as from a jogging load, the motor will deliver power to the electrical system briefly each time there is a voltage dip. This results in additional rotor currents as the flux re-establishes to the new conditions. When the voltage returns to

Table 14
Total Industrial Electrical
Consumption (1972) (billions kWh)*

Industrial motor drive (except HVAC†)		
Pumps	143	
Compressors	83	
Blowers and fans	73	
Machine tools	40	
Other integral hp applications	52	
DC drives	47	
Fractional hp applications	20	
Total drives		458
Other industrial electrical usage		
Electrolytic	—	
Direct heat	—	
HVAC*	—	
Transportation	—	
Lighting	—	
Total other		142
Total all industrial motors		600

*(Data from DOE, *Energy Efficiency and Electric Motors*, Conservation Paper 58, August 1976, p 26. Reprinted with permission.)

† HVAC is heating, ventilating, and air-conditioning equipment.

Table 15
Current and Future Motor Full-Load Efficiencies
in Integral hp AC Polyphase Motors

	Current			Future
	Worst	Best	Average	Improved Efficiency Models
1 hp	68	78	73	85.5
5 hp	78	81.5	80	89
10 hp	81	88	85	90
50 hp	88.5	92.0	90	92.5
100 hp	90.5	92.5	91.5	93
200 hp	94	95	94.5	95

*Data from DOE, *Energy Efficiency and Electric Motors*, Conservation Paper 58, p 43. Reprinted with permission.)

normal, there will be an additional current flow into the motor similar to a starting inrush. This will likewise cause additional rotor current to flow briefly with attendant losses.

A 1972 survey showed that motors comprised 75% of the electrical usage in industrial applications as shown in Table 14. Therefore, effort has been directed toward improving the efficiency

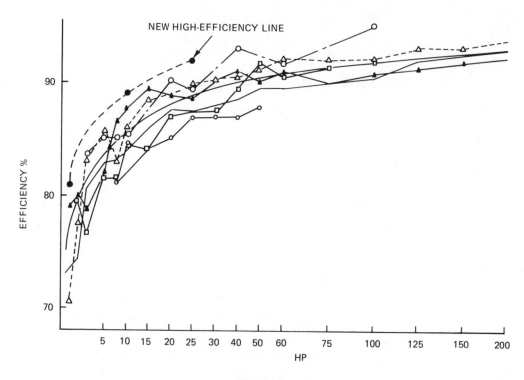

Fig 11*
Published Motor Efficiencies of
Principal Manufacturers
(Open, drip proof, 1800 r/min, NEMA Design B)

of motors. High-efficiency motors are now competitive with conventional types when consideraton is given to the cost of motor losses. Efficiencies of conventional and high-efficiency motors are illustrated in Fig 11 and Table 15.

5.2 Synchronous Motors. Synchronous motors have a more complex relationship with their electrical environment than induction motors. This is because of other interactions with the excitation system and the fact that imbalance and

distortion currents will flow in parts of the rotor circuit not generally designed for continuous current flow. Information on specific motors is generally available from the manufacturer.

5.3 Solid-State Devices. Diodes and thyristor-controlled rectifiers are frequently used to control the input power to motors and to nonrotating loads such as resistance heaters and chemical pot lines. Maximizing the energy utilization of these systems requires an understanding of diodes and thyristor characteristics and the efficiency characteristics of the loads they supply. A discussion of the load characteristics is beyond

*Data from DOE, Energy Efficiency and Electric Motors, Conservation Paper 58, p ES-9. Reprinted with permission.

the scope of this recommended practice since each is dependent upon a specific set of output parameters. A characteristic common to all diodes and thyristor loads is the distorted currents caused by their operation. Electrically, these loads act as *generators* of harmonic *currents*. The harmonic voltages, which are observed in the system, are the result of these currents flowing through an impedance. This is in contrast to a conventional voltage generator with resulting currents limited by the system impedances. In both cases, the conventional network equations are applicable, but the effect of a change in a given network constant will be significantly different. For example, increasing an impedance will reduce the current in the constant voltage case and increase the voltage in the constant current case. The maximum harmonics which can generally be created by one of these static conversion systems is given in Table 16. In practice, the harmonics will be less than shown in the table by 10%–30% unless resonance occurs.

Note that the harmonics are related to the alternating current which is related to the direct current. The dc voltage and,

**Table 16
Composition of Three-Phase
Alternating Current to a
Fully Smoothed DC Load**

Harmonic	Magnitude*
Fundamental	1.0 per unit (pu)
5th	$\frac{1}{5}$ = 0.200 pu
7	$\frac{1}{7}$ = 0.143 pu
11	$\frac{1}{11}$ = 0.091 pu
13	$\frac{1}{13}$ = 0.077 pu
17	$\frac{1}{17}$ = 0.059 pu
19	$\frac{1}{19}$ = 0.053 pu
N	1/N

*For a 6-pulse phase controlled rectifier.

hence, power level of the drive is not an indication of the harmonic content. Similar considerations will show that the power factor of a thyristor drive is a function of the output voltage and is not directly related to the load current.

$$PF = \frac{\text{average dc output voltage}}{1.4 \cdot \text{rms ac input voltage}}$$

The actual conversion efficiency of a thyristor drive will range from 90% to 98% for all loads above 25% of full load. This is shown in Fig 12.

**Fig 12
Thyristor Drive Characteristics**

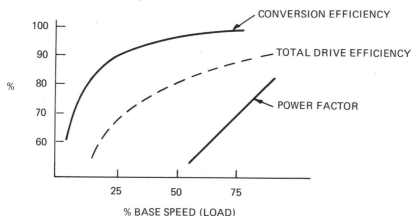

From an energy conservation viewpoint, thyristor drives should be operated at the highest possible power factor. With a given dc output voltage requirement, this can only be attained by lowering the ac supply voltage until the thryistors are phased full-on. The total energy required by such a drive also includes the system losses for the reactive component and the harmonics and the system losses for the real power component. As noted above, the harmonics will not be affected significantly by the operating voltage.

5.4 Transformers. Transformers seldom operate constantly at full load. Thus, their cyclic efficiency is more important than the rated efficiency. The specific cycle to which a transformer is evaluated should consider projected future conditions of operation and initial conditions. Many transformers shall be sized so that they can carry an extra load during emergency or routine outage conditions. This can result in selecting an oversized transformer for normal conditions in which case the transformer magnetizing losses become more important. The magnetizing (no load) losses are present

24 h a day for 365 days per year. When these overload conditions can be predicted, it may be desirable not to select an oversize transformer, but rather to deliberately consume a small percentage of the transformer life during the overload period. Figure 13 shows a typical relation between overload and duration of overload for no loss of life and for a 1% loss with an oil-filled transformer.

Transformers will have a reactive loss (I^2X) and a real power loss. When high reactance transformers are selected (for example, so as to minimize breaker short-circuit duties), the resultant increase in I^2X manifests itself as poor regulation of voltage. This may require additional voltage control with attendant increase in cost, complexity, and losses. Distorted voltage waveforms will also increase transformer losses.

5.5 Reactors. Reactors contribute losses which are largely reactive, although their real power losses can be significant when they are carrying rated current. One generally tries to arrange the system one-line diagram so that the reactors are located where they normally carry little current. The reactor size should not be

Fig 13
Transformer Load Versus Loss of Life

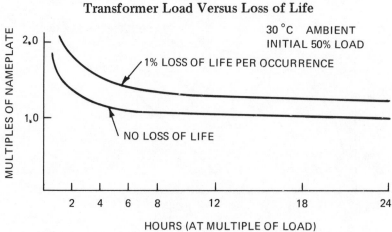

larger than needed to provide the desired electrical isolation. This will minimize voltage drops and heat loss which can affect adjacent equipment. The external fields of large air core reactors can induce currents in adjacent metal building components, and these losses are not included in the reactor specifications. Distorted currents will increase reactor losses. They also increase the hazard of a low-frequency resonance when capacitors are also used on the systems.

5.6 Capacitors. Capacitors have negligible losses under normal conditions. Their main contribution is to reduce losses in other equipment by providing the required reactive current locally, thus eliminating it from the supply lines. Capacitors are sensitive to wave distortion and care shall be used to minimize problems in this area.

When capacitors are used with thyristors or other nonlinear loads, a harmonic resonance between the capacitance and

system inductance can cause equipment damage throughout the plant. Even when resonance does not occur, harmonic re-enforcement can occur and this can lead to an increase in system losses greater than the reduction in fundamental frequency losses. Table 17 shows the increase of all harmonics below the resonant frequency. Harmonics substantially above the resonant frequency are reduced by the capacitor.

5.7 Frequency Effects. Circuit components and switchgear ratings are based on sine-wave voltage and current. Under these conditions, their losses are minimal and will vary with the square of the current. Wave distortions can increase these losses substantially, which will in turn increase the ambient temperature. This higher ambient can cause the direct acting (thermal) breaker trip mechanisms to operate on less current or in a shorter time. Additionally, these tripping devices are more sensitive to the higher frequencies which are in these distorted currents.

Skin effects have been mentioned several times as causing an increase in losses in all electrical equipment. Tables 18 and 19 show the skin effects on simple round conductors free of any magnetic material.

Proximity of magnetic materials may increase skin effect several times. Proximity of adjacent conductors will also increase skin effect, then called proximity effect.

Table 17
Increase in Harmonics Due to Capacitor

Harmonic Frequency as % of Resonant Frequency	% Increase in Harmonic Voltage
100	Q*
75	129
50	30
25	7
10	1

* Q is a function of system losses at the resonant frequency. The resonant frequency is given approximately by the formula:

$$f = 60 \frac{kVA\ (SC)}{kVA\ (CAP)}$$

where both kVA (SC) and kVA (CAP) are the symmetrical three-phase values for the system short circuit (SC) capability and capacitor size (CAP), respectively. This equation ignores circuit resistances and assumes that the capacitors are lumped at one electric location.

Table 18
Skin Effect on a 300 MCM Conductor

Frequency (Hz)	Harmonic of 60 Hz	AC/DC Resistance
60	1	1.01
300	5	1.21
420	7	1.35
660	11	1.65

Table 19
Skin Effects on Large Conductors

	AC/DC Resistance	
Approximate Round Conductor	60 Hz	300 Hz (5th Harmonic)
300 MCM	1.01	1.21
450 MCM	1.02	1.35
600 MCM	1.03	1.50
750 MCM	1.04	1.60

In general, where special shapes or conductor configurations are required for the desired 60 Hz capability, the skin effect may be considerably greater than given in Tables 18 and 19. While skin effect is growing in importance due to the increase in nonlinear loads, it generally is not a problem except where the dominant load on a substation bus is nonlinear or where power-factor capacitors are used extensively.

5.8 Size and Energy. The economy of size which pervades much of the electric industry planning is not generally applicable to an industrial installation. Here the size is dictated by the planned product output. However, there are some options which can lead to improved efficiency. Transformer efficiency increases with size. Thus, a plan which uses fewer, but larger transformers may be more economical from an operating and equipment investment viewpoint. Large motors are more efficient than small ones, but advantage of this fact cannot be taken generally. Switchgear and distribution losses per kilovoltampere are less at the higher distribution voltages. Thus, the higher voltage is generally preferred whenever the economics are marginal. Operating losses on the distribution system are generally less when the system is laid out in a network which results in a *working*

spare. The alternate philosophy of designing the system with an *idle spare* will result in increased operating losses but may result in a lower switchgear investment.

5.9 Voltage Considerations. The optimum solution to the reactive needs of the various components requires a complete system study. However, it is evident that the reactive needs are a function of both the voltage level and the power level of the various loads. Thus, a regulated voltage to each load can provide optimum voltage for minimum losses or minimum reactive needs, or both. While it would probably not be economical to regulate the voltage to each load, it is frequently feasible to group loads for voltage compatibility and to regulate the common service bus. This concept is in direct conflict with the principles of diversity which can lead to an improved sharing of facilities and, hence, less investment costs. In actual practice, the two concepts are not necessarily in conflict. The value of the diversity is greatest during an outage with a resulting overload on the remaining equipment. During such periods, load buses may be tied together and operated at a common voltage. But during normal periods, buses can be kept separate and at their respective optimum voltages. The total reactive need of a system is the diversified total required by the loads and the exciting reactive required of the transformers plus the I^2X of all of the distribution facilities. The reactive requirements of thyristor drives will have less diversity than the real power diversity. This is because the reactive current is a function of the direct current and with most of such drives working on a nearly constant current load, the reactive current will be substantially constant.

When the bus voltage is regulated for any reason it is generally accomplished with an on-load tap changing transformer (LTC) or an induction or step regulator. The resulting voltage will be a near replica of the supply sine-wave voltage. When the voltage to individual loads or motors is regulated, these same devices may be used or the equivalent of a lamp dimmer may be considered. When this latter solution is contemplated, the full impact of the resulting distorted voltage should be evaluated prior to the actual installation of such a device. The various performance parameters given in this section are based on sine-wave voltage. They should not be used to interpret the effect of voltage control wave distorting voltage regulating devices. This limitation is applicable even when the regulated voltage is expressed as a true rms value.

Excess voltage can result in reduced insulation life even though the operating temperature is held constant. This is particularly true for capacitors and other components which are worked near their corona threshold voltage. For capacitors the equation for insulation life is

$$L = \frac{1}{V^{7.45}} = V^{-7.45}$$

For a sustained 10% over voltage

$$L = \frac{1}{(1.1)^{7.45}} = 0.49 \text{ pu of rated life}$$

Thus, it seldom pays to deliberately operate capacitors or other equipment significantly above their rated value regardless of other factors which might favor such operation.

Momentary overvoltages may or may not be cumulative in their effect on loss of life depending on several factors. Generally, electrical devices that are equipped with surge arresters will have less frequent insulation failures in a hostile environment than other equipment not so protected. Surge suppressors may be applied for this reason.

5.10 System Evaluation. A system designed for maximum efficiency of energy utilization should also be the most economical system if the energy concept is viable. To determine the most economical system, operating costs, and first investment costs shall be treated over the full life of the facility. Section 3 dealt with this subject. A few additional points are pertinent in this section.

The cost of money will normally be a few percentage points more than the inflation rate. Thus, it is not economical to build in advance of need if one assumes a constant or falling rate of inflation over the life of the equipment. On the other hand, the cost of making an unplanned addition to a facility in the future can be very costly.

When the ampere rating of the switchgear is selected on the high side, the I^2R losses will be lessened. If the gear is outside an air-conditioned space, a significant reduction in air-conditioning operating costs can result. If the losses from the electrical equipment can be used to heat the conditioned space, a very-high energy utilization can be realized.

The energy charge for the losses may or may not be a small part of the total cost or value of these losses. Even the cost of ventilating an isolated switch room can be appreciable in terms of the billed cost of the losses.

Equipment in direct sunlight will operate less efficiently (hotter) than in a shaded area. The radiant energy of sunlight impinging on a 5000 kVA transformer is approximately equal to the full load losses of the transformer.

No utility presently charges for the distorted current required by thyristor drives beyond that inherent effect in the electromechanical meters. These may not accurately respond to harmonics. The trend is to electronic metering evaluation. Should this develop, the capability to meter the distortions will exist and charges for such distortions (and credits if they are forced on a user) may follow. It should be noted that, while existing billing meters largely ignore the distortions, the losses caused by the distortions beyond the metering point will be reflected through the nonlinear load as 60 Hz losses and will be registered on the billing meters.

The above comments apply to new and existing facilities. Existing facilities need additional comment especially when they are to be upgraded or expanded. In general, excitation losses are less in newly designed equipment. On the other hand, the newer equipment is more sensitive to voltage deviations. The excitation losses for most new motors and transformers increase more rapidly above nameplate voltage than the older equipment. Modern motors are rated 230 V or 460 V; motors built prior to 1970 may have a 220 V or 440 V nameplate. Regardless of any change in standards, operating both motors off the same supply bus will entail a lower overall efficiency.

A system optimized for energy conservation should also be an economically optimized system. When this does not appear to be the case, a review of the premises underlying the loss evaluation, operating and maintenance costs, and the economic treatment of the capital costs should be made. Also a reappraisal of the system one-line diagram may identify a situation which can be improved with a slightly different approach. For example, the cost of tap changing under

load (LTC) transformers may not appear warranted from information in hand on the utility voltage variations. However, LTCs provide the additional feature of regulating the load bus to maximize an output for one condition and minimize the need for reactive compensation under other conditions.

When evaluating an existing facility for operating cost reduction, a good starting point is the voltage levels on the various buses. While it is generally not economical to add tap changing under load equipment, it may be economical to switch capacitors to get a degree of bus voltage regulation. When capacitors are added for power-factor correction, the voltage will rise. This may unnecessarily increase equipment energizing losses and a manual tap change to a lower voltage may be desirable. On the other hand, if capacitors are needed only to increase the voltage, a transformer replacement with additional taps may be more economical.

Another area for evaluation is the practice of leaving unloaded motors on the line. Exciting energy costs may be more than formerly assumed. This is especially true if the voltage is high, as is not unusual now, or if there is serious wave distortion at the motor terminals. On the other hand, it is not economical to shut down a motor when not needed for too brief a period. When this is done, the energy stored in the rotating mass is lost and shall then be taken from the system during the restart. However, a motor is very inefficient during startup. Some motors can only tolerate two or three starts per hour or they will overheat. This energy loss plus the additional wear and tear on the motor contactor and motor from frequent stops and starts places a lower limit on the time limit a motor should be shutdown for minimum operating and maintenance cost.

Occasionally, one finds a practice of switching off transformers during light load periods with the objective of reducing the total exciting loss on the system. The increased load loss on the remaining transformer may be greater than the exciting losses saved if it is highly loaded at a high load factor.

When an existing facility is enlarged or upgraded to a higher capacity, some of the existing wasteful energy practices can be eliminated or minimized. Such an upgrading can best be started by listing all of the energy inefficiencies in the present facility regardless of whether they appear to relate to the proposed work. An analysis of this list may then suggest another way of proceeding with the upgrading which will be more cost effective and energy saving.

5.11 Energy Saving Devices. There are several *energy saving* devices on the market and more will undoubtedly appear from time to time. Any device claiming electric bill savings shall either reduce the need for energy or reduce the losses associated with the use of the electric energy. In either case before applying such a device, investigate to assure that:

(1) The claims are supported by scientific testing under controlled conditions with a base for comparison.

(2) The device is based on a scientifically supported phenomenon that is understandable by the user.

(3) The comparison of usages in a single plant or in each of a number of plants is not used because this method does not supply sufficiently controlled data to be meaningful.

This section has given information on voltage, phase balance, and other electrical environment effects on motors and other electrical equipment. A consultant or plant engineer should be able to evaluate any device by determining the theory of its operation and then testing this theory against the known operating characteristics of the equipment as noted in this section.

5.12 Bibliography

[1] ELECTRIC UTILITY ENGINEERS of Westinghouse. *Electric Utility Engineering Reference Book*, East Pittsburgh, PA: Westinghouse Electric Corporation, 1959.

[2] GIBBONS, William P. Analysis of Steady-State Transients in Distributed Low-Voltage Power Systems with Rectified Loads, *IEEE—IAS Transactions*, Jan/Feb 1980, pp 51-59.

[3] HICKOK, Herbert N. Electrical Energy Losses in Power Systems, *IEEE Transactions on Industry Applications*, vol 1A-14, no 5, Sept/Oct 1978, pp 373-386.

[4] KEY, Thomas S. Diagnosing Power Quality-Related Computer Problems, *IEEE—IAS Transactions*, July/Aug 1979, pp 381-393.

[5] LINDERS, John R. Effects of Power Supply Variations on AC Motor Characteristics, *IEEE Transactions on Industry Applications*, vol 1A-8, no 4, July/Aug 1972, pp 383-400.

[6] LINDERS, John R. Electric Wave Distortions, Their Hidden Costs and Containment, *IEEE Transactions on Industry Applications*, vol 1A-15, no 5, Sept/Oct 1979, pp 458-471.

[7] NOLA, Frank. Power Factor Controller — An Energy Saver, *Conference Record IEEE Industry Applications*

Society Annual Meeting 1980, Pub no 80CH1575-0, vol 1 of 2, pp 194-198.

[8] SHIPP, David D. Harmonic Analysis and Suppression for Electrical Systems Supplying Static Power Converters and Other Nonlinear Loads, *IEEE Transactions on Industry Applications*, vol 1A-15, no 4, Sept/Oct 1979, pp 453-457.

6. Metering and Measurement

6.1 Reasons for Metering. There are three important reasons for metering electrical energy in addition to the basic need to measure performance and status of the electrical system:

(1) To provide data for an energy audit.

(2) To allow proper distribution of electrical energy costs to individual departments based on periodic meter readings.

(3) To provide historical data on which to base electrical energy standards for the product produced or the service offered, and evaluate performance against that standard.

Audits based on a well-designed metering system will frequently yield surprising results and may often identify considerable savings. The results obtained from the audit will normally be in direct proportion to the effort expended. For example, in kilowatt demand control, many vendors gloss over the importance of a survey of plant loads to identify those that are suitable candidates for load shedding. In some cases, the vendors may acknowledge the need for such a survey but underestimate the time and manpower required to do a thorough job.

It is thus apparent that a formal, standardized survey method needs to be used if truly useful information is to be obtained. The survey sheet shown in Table 20 is typical and is used to tabulate all loads 5 hp (or 5 kVA) and larger. One sheet should be used for each common bus switching point or motor control center (MCC).

This survey form involves using current electrical drawings where available, supported by field checks to verify exact conditions whenever questions exist. Approximately one manhour, on the average, is required to complete each sheet, with some follow-up time to resolve occasional questions that arise when reviewing the survey results. Actual running kilovoltampere and kilo-

Table 20
Typical Form for Recording Electrical Load

Description and Equipment Number	Ownership	Connected kVA	Running kVA	Running kW	Cubicle	Operating Cost/Day	Estimate of Time (min) On	Off	Conditions Under Which Device Can be Down	Remarks (Utilization, Spares, etc)
Supply Fan K-7801.05		100			1D		10		For kWd control, or as part of	Consider as one load to be shed
Return air fan K-7801.06A		25			2C			50	cycling for kWd reduction	
Return air fan K-7801.06b		25			2D					
Elevator		75			2E		0		None	Not available
Air handling unit for S-3	S3	7.50			3D		55	5	kWd control only	Effects yarn quality
Service panel K-15		30			31L		0		None	Not available
Lighting panel K-78		30			31R		0		None	Review need for photo cells on some lighting circuits
Unit Substation KL-8	MCC K6-8-1					Date 11/18/81 Mo Day Yr	Prepared by WLS		L-5-4230-05 Ref Drawing Number Page 1 of 1	Rev 0

NOTE: Surveying plant loads is essential to know where energy is used; it is essential to survey plant loads. This typical form is for recording electrical-load data. Similar forms can be used for each motor control center in the plant and for all loads of 5 hp or 5 kVA or more. On an annual basis to keep the data current approximately 1 man-hour per form is needed.

watt data are obtained using a variety of electrical meters described later in this section.

Typical survey results can be presented in many forms; two examples are shown in Table 21 and Fig 14. Table 21 lists electrical system usage by function for a typical industrial plant and points out the major users. In this example, approximately 34% of the total load is associated with the production and distribution of chilled water for air conditioning.

Table 21
Typical Electrical System
Usage by Function for a
Typical Industrial Textile Plant

	%* of Total Load
HVAC Fans†	13
Chillers	12
Compressed air	11
Texturizing	8
Extruding and metering	6
Quench air	6
Cooling water pumps	5
Staple spinning	5
Lighting	5
Filament spin draw	4
Chilled water pumps	3
Chip drying	3
Staple drawing	3
Polymerization	2
Filament draw twist	1
Staple tow drying	1
Filament spinning	1
Waste treatment	1
Nitrogen, inert gas	1
Cooling tower fans	1
Beaming	1
Miscellaneous	7
Total	100.0

*Percent values are rounded to the nearest whole number.

†Heating, ventilation, and air conditioning.

A second approach is illustrated in Fig 14 using block diagrams to detail the efficiency of the plant electrical system. Data for this diagram is collected using portable electrical meters, combined with the data form shown in Table 20.

Note in Fig 14 that the overall plant efficiency is only approximately 81%, which means that 19% of all electrical energy is wasted as heat. This may require additional air conditioning to remove the heat from the production areas. From this typical diagram, it appears that effort should quickly be directed toward improving the 53.3% efficiency of the variable frequency drive systems which use 7% of all the plant's electrical energy.

With regard to audits of the electrical power system, metering results can be used to compile an energy profile that can:

(1) Aid in establishing and refining energy use by product line, department or area.

(2) Establish and improve energy-use accountability.

(3) Allow measurement of cost reductions.

(4) Help determine equipment capabilities and load factors for future modifications and plant expansion.

(5) Provide data for analyzing results that vary from established standards.

In addition, an energy profile likely will reveal areas where conservation projects are most beneficial.

The second reason for metering involves the need to accurately distribute the charges for energy to each department on a periodic basis. In times past, only a single kilowatthour meter may have been available, usually at the utility company metering point. There was no attempt to subdivide the energy based on further meter readings. Today, how-

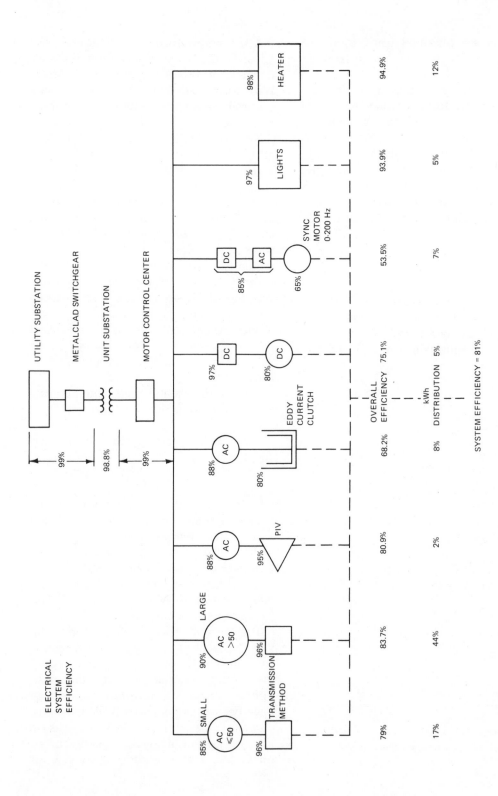

Fig 14
Portraying Plant Efficiency

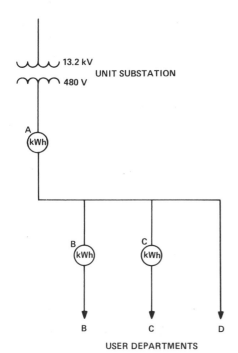

NOTE: One way to hold down costs is to measure consumption in some departments by taking the difference between two meters. Although some accuracy is sacrificed, only three meters are needed to obtain four values.

Fig 15
Subtractive Metering

ever, the trend is to provide sufficient metering to completely identify the users and allow charges to be made to individual departments. Once a metering system is installed and operating, a 1% energy savings can be expected as a result of the detection of system losses.

Metering can be costly. The cost of installing kilowatthour meters is usually between $300 and $1000.[13] One way to keep costs at a minimum is to measure consumption in certain departments by taking the difference between two meters. Figure 15 illustrates the con-

---

[13] Estimate determined in US dollars in 1980

cept. Meter A measures the total amount of electrical consumption through the unit substation. Meters B and C measure consumption by Departments B and C. Consumption by Department D is determined by $D = A - (B + C)$.

Some typical problems in establishing a first class metering system include:

(1) Insufficient funds to monitor as many circuits as desired

(2) Poor location of conductors that do not lend themselves to proper Potential Transformer (PT) and Current Transformer (CT) device locations

(3) Loads that cannot be shut down to permit installation of proper metering devices

(4) Meters in hard-to-reach or dangerous places

(5) Engineers unfamilar with what is needed, and lack of trained calibration and maintenance personnel

(6) Inadequate funds for maintenance and spare parts

(7) Lack of time or qualified staff to read the meters

(8) Meter readings that do not reconcile (for example, meters that indicate more energy was consumed than was purchased from the utility company).

The third reason for metering involves gathering historical consumption data on which to base electrical standards and then measure performance against that standard. For example, suppose an electrical standard of 1 kWh/lb of product has been determined for a given department. During the first week, that department uses 1000 kWh to produce 1000 lb. The actual usage ratio is:

$$\frac{1000 \text{ kWh}}{1000 \text{ lb}} = 1 \text{ kWh/lb}$$

The value is plotted at day 7 in Fig 16 and represents the expected, satisfactory electrical performance.

The second week the department im-

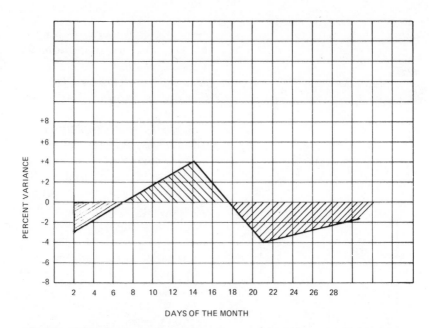

NOTE: Comparing actual weekly consumption versus production to a standard value for each process allows the progress of the program to be evaluated easily. Ideally, more plot points should fall above the standard line than below. It is always desirable for the slope of the line between the plotting points to trend upward.

Fig 16
Weekly Consumption Versus
Production Compared to Standard

proves and uses only 960 kWh to produce 1000 lb. Now, the actual usage ratio is:

$$\frac{960 \text{ kWh}}{1000 \text{ lb}} = 0.96 \text{ kWh/lb}$$

The variance is calculated:

$$\frac{\text{standard} - \text{actual}}{\text{standard}} \cdot 100\% = \text{variance}$$

$$\frac{1.0 - 0.96}{1.0} \cdot 100\% = \frac{0.04}{1} \cdot 100\% = +4\%$$

This better-than-average value is plotted at day 14. The third week the department becomes careless and uses 1040 kWh to produce 1000 lb. Now, the actual usage ratio is:

$$\frac{1040 \text{ kWh}}{1000 \text{ lb}} = 1.04 \text{ kWh/lb}$$

The variance is calculated:

$$\frac{1.0 - 1.04}{1.0} \cdot 100\% = \frac{0.04}{1} \cdot 100\% = 4\%$$

This worse-than-average value is plotted at day 21.

By the end of the month, more plot points should fall above the standard line than below. The absolute value for any given week is not as significant as the slope of the curve between the weekly plotting points. It is always desirable for the variance curve to trend upward. Plot points that are always below the zero variance line may indicate poor performance, but it may also mean the standard is too stringent. Similarly, points that always lie above the zero variance line may indicate superior performance or too lax a standard.

Taking meter readings on a more frequent basis, perhaps as often as each shift, can reveal useful information such as:

(1) Meter misreadings

(2) Meter error

(3) Load changes due to production changes

(4) Most and least efficient shifts

(5) Need for maintenance

6.2 Metering Energy Flow. Metering of power demand and energy consumption in the primary distribution system, generally considered to be at the 5 kV voltage level and above, consists of conventional CT and PT supplied voltmeters, ammeters, and kilowatthour meters. Specialized equipment includes combinations of these functions that provide kilowatthour, kilovarhour, kilovoltampere hour, kilowatt, kilovoltampere, and instantaneous power-factor data by means of dials and a strip chart recording.

Other specialized equipment includes the so-called *demand recorder* which may be as simple as a dial pointer attachment on a kilowatthour meter or as complex as a microcomputer-based data logger, printing kilowatt demand and kilowatthour consumption on a continuous strip of paper at periodic intervals of 15, 30 or 60 min. Data from these recorders indicate exactly when the peak occurred and can be used in conjunction with the energy survey and audit to identify the primary causes of the peaks. These recorders are generally connected to pulse contacts supplied by the utility company for a monthly fee, or may be operated from the customer's own PTs and CTs.

While the data loggers cost from $1000 to $3000,[14] it is not uncommon for

them to pay for themselves in less than six months by providing data and alarm points for use with a manual load shedding program. Table 22 provides approximate costs (1979) for several levels of complexity.

Metering of loads below 5 kV within a load center is difficult to justify due to the relatively high initial cost. However, these instruments can often be justified by potential savings in operating and maintaining the plant's electrical equipment. Checks can be made periodically with portable instruments where permanent meters cannot be justified.

Regardless of where the meters are used, meter data shall be recorded and filed in a manner that will encourage review and reaction to changes. A well-designed log sheet for electrical consumption (Fig 17), in kilowatthours, should have at least these column headings:

(1) Date

(2) Present meter reading

(3) Previous meter reading

(4) Difference

(5) Meter multiplier

(6) Consumption

(7) Peak kilowatt demand reading since it was reset

Monthly consumption should equal the sum of all Column F readings. The sheet should be turned in to a central recording center to plot consumption in each department.

6.3 Basic Meter Components. The three basic types of meters are the voltmeter, ammeter, and wattmeter. Their typical connection diagrams are shown in Fig 18.

The voltmeter is a high-resistance device and is always connected in parallel with a source of power or load. The ammeter measures the current flowing through a conductor. It is a low-resistance device and is always connected

---

[14] Estimate taken in 1980 and reflects US dollars.

Table 22
Data Loggers and Demand Controllers

Category	Controller Cost Only	Installed Cost Per Control Point	Maximum No of Output Control Points	Program Changes By	Features
I Manual surveillance	$1000 to $4000	$1000 to $4000	1 (If the alarm point is used)	None available	Alarm and hard copy monitoring of kWd load
II Hardwired	$3000 to $15 000	$ 300 to $1000	16 to 48	Operating personnel using plugs or switches	Alarms kWd control kWh conserved cycling
III Minicomputer	$20 000 to $80 000	$ 200 to $ 800	48 to 640	Operating personnel using TTY CRT console, paper tape or function keys	Alarms kWd control kWh conserved cycling optimization logs graphs
IV Total energy control system	$100 000 to $400 000	$ 100 to $ 600	Several thousand	Operating personnel using TTY, CRT console, paper tape, or function keys	All of Category III above plus complete monitoring and control

A listing of data loggers, Category I; demand controller, Categories II and III; and complete energy control systems, Category IV.

in series with the source of power or load. The wattmeter measures apparent, real, or reactive power flow from the source to the load. The potential coils are connected in parallel and the current coils are connected in series with the load.

In cases of portable instrumentation, the ammeter and the current coils of the wattmeter usually obtain their current signal from a clamp-around current transformer. This current transformer (CT) is used to step down line current to a level that can be conveniently metered, generally to 5 A or less. The CT surrounds the primary conductor and produces a secondary current proportional to the magnetic field created by the primary current in the conductor being measured. The ratio of line current to secondary current is known as the CT ratio or CTR. CTs are normally rated in values to 5 A such as 100:5, 1000:5, and 5000:5. These values indicate how many amperes flowing in the primary conductor will cause 5 A to flow in the secondary winding.

Figure 17
Electrical Reading Log Sheet

Meter Name _____   Area _____

Location _____   Month/Year _____

Dept No _____   P.P.R.* _____

Reader Initials	(A) Date	(B) Previous Reading	(C) Present Reading	(D) Difference (C) – (B)	(E) Multiplier	(F) SCF|Consumed (D) × (E)	(G) Demand Reading Reset to Zero
Total Month (kWh)							

A typical log sheet for electrical energy consumption. The log is turned in to a central recording center and used to plot consumption in each department.

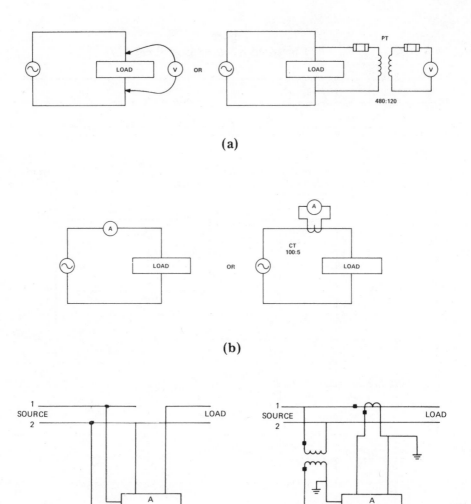

(a)

(b)

(c)

Fig 18
(a) Voltmeter (b) Ammeter
(c) Wattmeter

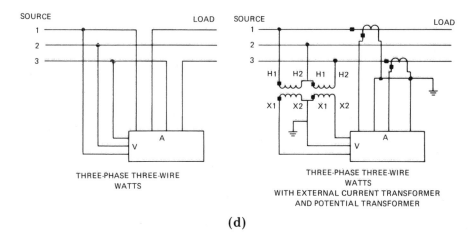

(d)

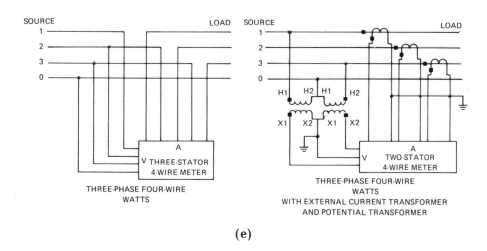

(e)

NOTE: Blondel's Theorem states that if a network is supplied through N conductors, the total power is measured by summing the reading of N wattmeters so arranged that a current element of a wattmeter is in each line and the corresponding voltage element is connected between that line and a common point. If the common point is located on one of the lines, then the power may be measured by $N-1$ wattmeters. This means a three-phase, three-wire system requires only two single-phase wattmeters or one polyphase instrument with two measuring elements. Four-wire circuits require three wattmeters or a three-element instrument.

Fig 18
(d) Three-Phase Three-Wire Wattmeter Connection
(e) Three-Phase Four-Wire Wattmeter Connection

The secondary of a CT shall always be a complete circuit whenever there is current flowing through the primary conductor. Thus, the leads of a CT shall never be fused and shall always be either connected to a low-resistance ammeter movement or shorted together by means of a jumper wire, screw, or switch on a CT shorting terminal strip.

A voltage step-down device known as a potential transformer (PT) is used to reduce line voltage to a level to match the meter rating, generally 120 V. The ratio of primary to secondary voltage is known as the PT ratio or PTR. The leads on a PT shall always be fused and shall never be shorted together. Voltage transformers are available in many accuracy classes, and it is important to select one suitable for a specific application. The two most important specifications for these transformers are ratio accuracy and phase-angle error. Phase-angle errors should be less than 50 min for use with analyzers of the 99% accuracy class and 10 min or less when used with instruments having accuracies of 99.75% or better.

The final accuracy of the measurements is in the ±3% range if the individual components, CTs, PTs, and meters are each ±1% accurate. It should be noted that control transformers are not suitable for metering since their voltage can be off by 10%.

The standard industrial wattmeter has 120 V rated potential coils and 5 A rated current coils, and is thus designed to work with appropriate PTs and CTs. The meter multiplier used to obtain actual power on the primary side is the meter reading times the PT ratio, times the CT ratio. This multiplier is usually shown on a small tag attached to the meter face.

Apparent power, kilovoltampere, is the product of volts times amperes. Real or true power, kilowatt, is the volts times amperes times the cosine of the phase angle between the voltage and the current. Reactive power, kilovar, is the volts times amperes, times the sine of the phase angle between the voltage and current. Meters reading reactive power obtain the correct signals by means of an external phase shifting transformer. The $\sqrt{3}$ shall be included when calculating three-phase power.

6.4 Meter Selection. A wide variety of meters are available to meet almost every measuring need. Basically, meters are divided into two types, indicating and recording.

Indicating types are used to provide a measured value at a given moment in time, and will show how the instantaneous value changes as a function of time. Indicating meters may have analog pointers or digital readouts. Some meters have a built-in time delay. Many are equipped to show average usage over a 15 min to 30 min period.

Recording meters are used to accumulate a measured value over a period of time. These meters use either dials such as a kilowatthour meter, a paper chart (either circular or in a strip), or magnetic tape where readings are *coded* into pulses and then uncoded by a computer system to record the increase in reading between reading intervals.

Power factor and negative sequence meters are available to meet specific needs. A power-factor meter indicates either a single or three-phase value leading or lagging. It requires both potential and current connections much the same as a wattmeter. Most power-factor meters are single-phase devices and are not accurate on either imbalanced voltage or current on a three-phase system.

Some of the new electronic meters respond to zero crossings and in the presence of distortion are not accurate. A negative sequence meter indicates the presence and magnitude of negative sequence voltage or current, usually associated with system imbalance or fault conditions, or both.

6.5 Meter Timing for Kilowatt Measurement.

The disc of a watthour meter makes from one-half to one rotation per watthour depending on the meter constant Kh. The number of watthours for any period is then the product of the number of disc rotations times the meter constant. The speed of rotation of the disc, therefore, indicates the usage rate or the watts being used. The kilowatt is a more reasonable quantity for consumption rate and the following formula can be used when timing a meter disc:

$$\text{Kilowatts} = \left[\frac{(PTR \cdot CTR) \cdot (Kh \text{ Wh/rev}) \cdot (3600 \text{ s/h})}{1000 \text{ W/kW}} \right] \cdot \left(\frac{\text{REV}}{\text{s}} \right)$$

$$\text{Kilowatts} = 3.6 \cdot PMC \cdot \left(\frac{\text{REV}}{\text{s}} \right)$$

where

- PTR = potential transformer ratio
- CTR = current transformer ratio
- Kh = meter constant found on the face of the meter (typically 1.2 to 1.8)
- REV = number of disc revolutions during observation period
- SEC = period of observation in seconds
- PMC* = primary meter constant = $Kh \cdot (PTR \cdot CTR)$

---

*This value is often stamped on the meter face plate and identified by the symbol PKh.

6.6 Determining Induction Motor Loads.

The detection and changeout of large underloaded induction motors to smaller or higher-efficiency induction motors, or both, will contribute greatly to improved system efficiency and power factor. The underloaded motors can be detected by the use of a portable torque analyzer. This device is basically an optical tachometer used in conjunction with electric meters and calibrated to read out directly in percent of shaft horsepower to the load. The principle of operation is based on the fact that the slip r/min of an induction motor is linear from 10% load to 110% load.

The greatest benefits of the torque analyzer or similar devices such as optical and mechanical tachometers include

(1) Ability to quickly locate underloaded induction motors.

(2) Ability to watch loading of equipment versus other conditions such as through-put, filter conditions, temperature, and pressure.

(3) Ability to determine true motor efficiency at any load when used in conjunction with conventional kilowatt monitoring on the motor input leads.

(4) Ability to assist in sizing future motor requirements based on actual load data.

(5) Ability to assist maintenance mechanics in periodically checking motor loading which helps detect worn bearings, clogged filters, etc.

There are many opportunities for engineering new devices to measure the efficiency levels in a plant. The engineer should combine his knowledge of metering equipment including oscillographs and other more sophisticated devices with his knowledge of external motor characteristics such as slip versus load, power factor versus load, etc. Devices similar to the torque analyzer can be

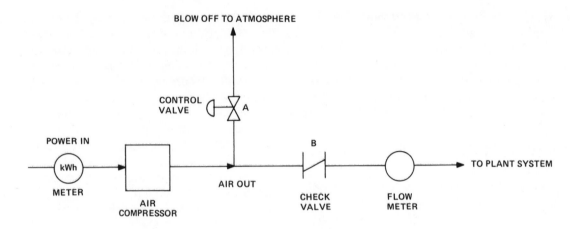

Fig 19
Method of Metering Air Flow from a
Centrifugal Compressor

effectively utilized with careful consideration of meter accuracy, voltage imbalance effects, and voltage level effects that are discussed in Section 5.

6.7 Practical Examples. The following examples are intended to serve as typical guides only and should not be interpreted as exact models for measurement.

Occasionally, the metering system will provide direct savings that can be immediately identified. For example, Fig 19 illustrates a method of metering air flow from a centrifugal compressor. Under normal operation, check valve B is open and control valve A is closed. Power into the motor and air-flow out of the motor are metered. Normal (kW/1000 ft^3)/min values are approximately 200, which means that 200 kWh are required for each 60 000 ft^3 of compressed air produced at a rate of 1000 ft^3/min. When the plant load decreases, the outlet pressure of the compressor increases, causing check valve B to close and con-

trol valve A to open. Air is vented into the atmosphere to prevent surging. An alarm light notifies the utility operator that the compressor is in the blow-off mode and needs to be shut down until the demand for air increases.

Over a two-week period, the daily kW/1000 sft^3/min values increased. When the problem was investigated, control valve A was found to be stuck in the open position and check valve B was closed. The alarm light had burned out and the compressor was venting air to the atmosphere. Left uncorrected, the situation would have wasted $9600 per month for electricity at 4¢/kWh.

A second example involves a similar monthly plot of electricity. The plot indicated that, at the end of the month, consumption would exceed the standard by 12%. The situation was corrected by shutting down four 100 hp air-handling units that, when running, consumed approximately $4000 worth of electricity a month. A week later, the updated plot still indicated consumption was too high and, according to the weekly electrical log sheets, consumption had not dropped on the substation that powered the four air-handling units that had been shut down. An investigation revealed the units had been restarted. Now, the shutdown procedure has been revised to lock out the unit and tag it as being shut down for energy management.

A third example involved monitoring of power to a 1500 ton chiller with a 2300 V drive motor. A conventional kilowatthour meter was used to record input power, and conventional input and output dial thermometers combined with a flow measuring device were used to calculate the output chilled-water tonnage rate. Instantaneous power was determined by timing the rotating disk (see 6.5) and using the following formula:

$$kW = 3.6 \cdot PMC \cdot \left(\frac{REV}{s}\right)$$

Initial calculation of kW/t, a fundamental indicator of chiller performance, resulted in values ranging from 0.28 to 0.34 kW/t. These values were only one-half of the 0.6 and 0.7 kW/t values usually associated with chillers of this size. A check of the kilowatthour meter and associated CTs and PTs revealed no unusual values. Attention was then given to the dial thermometers, which were found to be out of calibration and giving a false temperature differential. This resulted in an inflated tonnage rate value, causing the kW/t ratio to be unusually low. Replacement of the dial thermometers with a mercury-in-glass style corrected the problem and provided correct data on which to base future calculation of chiller performance.

6.8 Other Considerations. More complex methods than those that have been described can be used if special accuracy is desired, but it is usually unnecessary in a general survey. Moreover, the more complex the procedure, the more costly the equipment and the higher the operator-skills requirements. Higher precision may be desired if the survey uncovers areas where motor changeouts could save significant amounts of energy by sizing motors to run closer to their rated loads.

It is possible to be lulled into a false sense of security by assuming that everything is functioning as planned. For example, some meters may be recording properly while others are either running high, low, or slow. Therefore, it is wise to interrogate the system and people. For example, are you sure that the wiring is correct? How did you verify it?

Generally, kilowatthour meters run well when they are connected correctly. Faulty readings are caused by (1) current transformers and voltage transformers with reversed polarity connections, (2) voltage transformers with fuses blown, and (3) shorting blocks with shorting screws not removed.

The form shown in Fig 20 illustrates one method of performing a kilowatthour verification test. The instantaneous kilowatt power is read on each of up to six feeders using a clamp-on type meter. The sum of these readings is compared to the instantaneous kilowatt value obtained from timing the main kilowatthour meter disk. The values should be within ± 10%, if the load is relatively steady during the test interval.

6.9 Conclusion. The ultimate goal of a metering system is to assist plant personnel in managing the energy required to produce a specific product or service.

New Meter Verification Test

Unit Substation _____ Date: _____

Location _____ By: _____

Meter Factors:

    MM _____ MM _____ PTR _____ CTR_____

    Meter Multiplier _____ Name _____

    Style Number _____

Test Data Using Epic Model Clamp on kW Meter

	Meter Description	Volts	Amperes	Epic Meter kW Reading
1.				
2.				
3.				
4.				
5.				
6.				

Total MW _____

Main Meter Disk

    Seconds/One Revolution _____

Check: $\dfrac{\text{Total MW (PMC (3.6)}}{\text{s/Rev}}$

Fig 20
Form Used to Verify that Main kWh Meter is Reading Correctly
Compare to Individual Feeder Loads

Monitoring and reporting energy consumption allows for close control while minimizing expenses. Another goal is to provide historical energy consumption data to aid in projecting future loads and developing standards for the next year. Such data are essential to financial forecasts and operating budgets.

Unless energy consumption is measured, it is next to impossible to know where to direct conservation efforts. A metering system provides that vital ingredient to a successful energy management program.

The following list of check points should be used to make an effective energy survey:

(1) Make detailed plans prior to the actual survey.

(2) Use prepared forms so records can be easily compared on a year-to-year basis.

(3) On constant load where efficiency can be determined in a short time, consider using analog type indicating devices.

(4) Where loads fluctuate or have varying duty cycles, use an automatic logging system or chart-type recording devices.

(5) Check new equipment when installed to determine if it meets specifications and to provide data for reference in future surveys.

(6) Be sure to make the survey when equipment is in use at its normal production load.

(7) Make sure equipment has reached normal operating conditions: correct temperature, speed under normal load, etc.

(8) Have test equipment neatly arranged on a portable cart and be sure to include extra test leads, equipment fuses, etc.

(9) Make certain that personnel doing the testing are qualified to work on energized electrical equipment even if tests are done while equipment is de-energized and have the necessary safety equipment such as safety glasses, rubber gloves, and hard hats.

7. Energy Conservation in Lighting Systems

7.1 Introduction.

While far from the greatest user of energy, lighting is significant because it enters visibly in virtually every phase of modern life. The opportunities for saving lighting energy require that attention be paid to many nonelectric parameters.

Lighting systems are installed to permit people to see. Attention shall be paid not only to economics and efficiency but also to the type of work which people do and the space in which they do it. Lighting also affects other environmental and building systems, especially those which heat and cool occupied spaces within buildings. All of these factors shall be integrated into the design process.

This section presents the state-of-the-art in optimizing the use of lighting energy and it also alerts the lighting designer and user to the problems involved in designing an energy-efficient lighting system.

7.2 Definitions

ballast. A device used with an electric-discharge lamp to provide the necessary voltage and current for starting and operating the lamp.

candlepower. Luminous intensity expressed in candelas.

coefficient of utilization (CU). The ratio of the amount of light received on the work-plane to the amount of light emitted by the lamp.

color rendering. A general expression for the effect of a light source on the color appearance of objects compared with color appearance under a reference light source.

general lighting. Lighting designed to provide a substantially uniform level of illumination throughout an area, exclusive of any provision for special local requirements. *See:* **task lighting.**

glare. The sensation produced by brightnesses within the visual field that are sufficiently greater than the luminance to which the eyes are adapted to cause annoyance, discomfort, or loss in visual performance and visibility (see 7.7 for more detail).

illumination: A condition where light impinges upon a surface or object.

lamp. A generic term for a man-made source of light.

lamp lumen depreciation factor (LLD). The factor used in illumination calculations to quantify the output of light sources at 70% of their rated life as a percentage of their initial output.

louver. A series of baffles used to shield a source from view at certain angles or to absorb unwanted light. The baffles usually are arranged in a geometric pattern.

lumen (lm). A unit of luminous flux.

luminaire. A complete lighting unit consisting of a lamp or lamps together with the parts designed to distribute the light, to position and protect the lamps, and to connect the lamps to the power supply.

luminaire dirt depreciation factor (LDD). The factor used in illumination calculations to relate the initial illumination provided by clean, new luminaires to the reduced illumination that they will provide due to dirt collection at a particular point in time.

luminance. A measure of the amount of light flux (lumens) per unit of area reflected from or transmitted through a surface.

luminance contrast. The relationship between the luminances of an object and its immediate background.

luminous efficacy. A measure of lamp efficiency in terms of light output per unit of electrical input expressed in lumens per watt.

polarization. The process by which light waves are oriented in a specific plane.

reflectance (of a surface or medium). The ratio of the reflected flux to the incident flux.

shielding angle (of a luminaire). The angle between a horizontal line through the light center and the line of sight at which the bare source first becomes visible.

task lighting. Lighting designed to provide illumination on visual tasks usually at higher levels than the surrounding area.

7.3 Method of Presentation. In the succeeding sections, the major elements of the lighting system will be analyzed. For more detailed information the reader should consult the appropriate reference in the Appendix.

A lighting system is that portion of the branch-circuit wiring system which supplies the lamps or ballasts together with the associated controls such as switches and dimmers. The system also includes the light source(s), luminaire, shielding, and optical control media, the entire space to be lighted, and the nature of the illumination required. The effectiveness of the entire system is expressed in *figures of merit*. A generalized diagram of this *system* and the associated figures of merit are given in Fig 21. Specialized lighting terms used in this section are briefly described in 7.2.

The human eye is not equally sensitive to all wavelengths in the visible spectrum. Colors in the green and yellow region

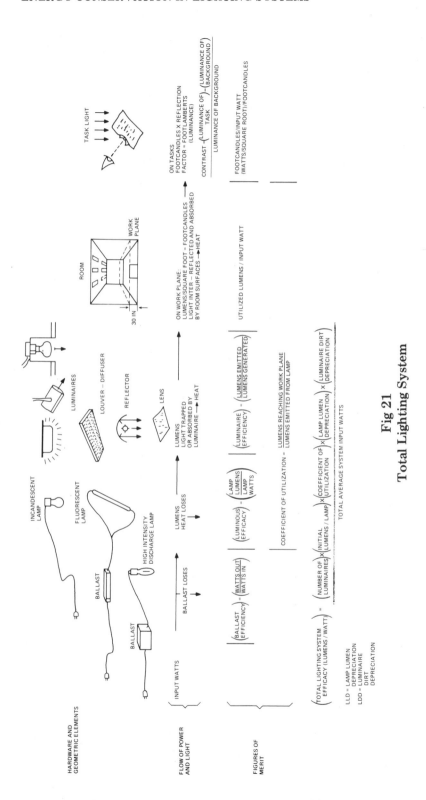

Fig 21
Total Lighting System

will produce a sensation of brightness in the eye with less radiant power than will wavelengths in the blue and red region of the spectrum. Lighting units or *photometric* units take this sensitivity response into account in their definitions. Photometric quantities therefore apply to human visual response rather than the strictly radiant effects of energy.

7.4 The Task and the Working Space

7.4.1 Task Description. Usually a visual task is considered the *given* in a lighting system and the design process starts from there. However, optimizing energy for lighting requires that consideration be given to the possibility of changing and improving the visual task characteristics so that the lighting requirements of the task are less stringent.

Vision requires sufficient light (brightness) and size for the eye to resolve an image, sufficient time to recognize an image and contrast to separate the brightness of an object from its background.

7.4.2 Task Illumination. A century of research information has resulted in tables of recommended illumination values for different visual tasks, scaled to reflect the accuracy with which the task can be performed. Improved instrumentation and understanding of the visual performance process are resulting in better methods of determining optimum task illumination. Measures such as task difficulty and criticalness are being combined with characteristics of the human seeing mechanism, including such variables as age and eye defects.

Most lighting recommendations are given for *illumination on the task* and presume that there is little need to illuminate the entire space to the task-illumination requirements. Recent studies involving work stations rather than general lighting have shown that

under certain conditions the energy requirements for lighting spaces can be substantially reduced without sacrificing task visibility.

Generally, improving the reflectance characteristics of the task and the quality or the quantity of its illumination will improve contrast. The contrast is defined as follows:

$$\text{Luminance contrast} = \frac{\left(\begin{array}{c}\text{background}\\\text{luminance}\end{array}\right) - \left(\begin{array}{c}\text{object}\\\text{luminance}\end{array}\right)}{\text{background luminance}}$$

$$\text{Luminance} = \text{illuminance*} \cdot \left(\begin{array}{c}\text{reflection}\\\text{factor}\end{array}\right)$$

*Illuminance is the preferred term for illumination.

Contrast improvement may not be simple where other than matte objects and backgrounds are involved. Even with matte surfaces, contrast may be subtly reduced by a phenomenon known as veiling reflections, a condition where contrast is lost because of the reflection of the light source in portions of the task. Contrast losses may be large even on ordinary tasks such as reading printed words on paper.

Design and evaluation techniques are available which can improve task contrast by reducing veiling reflections. At present, calculation techniques involve the determination of *equivalent sphere illumination* (ESI). Illumination (footcandles) is referenced to a known task illuminated by a light source consisting of a perfectly diffusing hemisphere placed over the task. ESI calculations are referenced to sphere illumination to compare whether a given lighting system is worse or better than sphere illumination in producing contrast.

7.4.3 Efficient Room Lighting Design.
Since the room acts as part of the lighting system, careful attention should be paid to how light is directed from the lighting equipment to the task areas. The controlling factors in this process are the size and shape of the space, the reflectances of the surfaces, and the characteristics of the room furnishings. Light may be trapped, absorbed, reflected, or modified by any surface within the space. Substantial light may also be allowed to enter or escape by means of entrances and windows.

While the most energy-efficient room may result from the use of highly reflecting surfaces and furnishings, such a space may be psychologically and aesthetically objectionable. Trade-offs are necessary. Typical recommendations for room surface reflectances are shown in Table 23.

It is also important to consider the luminance of the surfaces surrounding the visual task to avoid great changes in contrast which can cause eye fatigue and discomfort. Keeping within recommended luminance ratios is particularly important where localized or task lighting is used since lighted areas are more concentrated and general room illumination may be substantially below task levels. Table 24 summarizes recommended ratios which may be useful in overall planning.

Table 23
Recommended Surface Reflectances for Offices

Surface	Reflectance Equivalent Range
Ceiling finishes	80%–90%
Walls	40%–60%
Furniture	25%–45%
Office machines and equipment	25%–45%
Floors	20%–40%

Table 24
Recommended Luminance Ratios

To achieve a comfortable balance in the office, it may be desirable and practical to limit luminance ratios between areas of appreciable size from normal viewpoints as follows:

1 to $\frac{1}{3}$ between task and adjacent surroundings
1 to $\frac{1}{5}$ between task and more remote darker surfaces
1 to 5 between task and more remote lighter surfaces

These ratios are recommended as maximums; reductions are generally beneficial.

7.5 Light Sources
7.5.1 Light Source Efficacy.
Table 25 shows the approximate allocation of

Table 25
Lamp Energy Data
(Nominal Data in %) Initial Ratings

Lamp Type	Radiated Energy Light	Infrared	UV	Conducted and Convected	Ballast
Incandescent (100 W)	10	72	—	18	—
Fluorescent (40 W)	20	33	—	30	17
Fluorescent (40 W excluding ballast)	24	40	—	36	—
Mercury (400 W deluxe white)	15	47	2	27	9
Metal halide (400 W clear)	21	32	3	31	13
High-pressure sodium (400 W)	30	35	0	20	15
Low-pressure sodium (180 W)	29	4	0	49	18

Table 26
Lamp Lumen Efficacies

Source	Wattage	Approximate Lumens/Watt Input*
Incandescent	40 W general service	11
Incandescent	1000 W general service	22
Fluorescent	2 24 in coal white	50
Fluorescent	2 48 in cool white	70
Fluorescent	2 96 in (800 mA) cool white	73
HID	400 W phosphor-coated mercury	50
HID	1000 W phosphor-coated mercury	55
HID	400 W metal halide	75
HID	1000 W metal halide	85
HID	400 W high-pressure sodium	100

*Lamp only.

energy use in various light sources. A light source can be characterized by light output, life, maintenance of light output, color, etc. A more useful measure is lamp luminous efficacy which is the amount of visible light emitted by the lamp per watt of input power. As Table 26 indicates, the incandescent lamp has the lowest efficacy at between 10–20 lumens per watt, and mercury is approximately double that at about 50 lumens per watt. Lamp efficacies may improve as new lamps are developed.

While lamp efficacy is a good indicator of the lamp's ability to change electric energy into visible energy. It does not include the losses which take place in the luminaire and the room as the light is directed toward the task area. A true measure of system efficacy should include the utilization characteristics of the fixture, the room and the effect of dirt cumulation. These loss factors may be light source dependent, therefore using the most efficient lamp may not always result in the most efficient lighting system.

7.5.2 Light Source Characteristics. The light source shall match the environmental constraints. For example, fluorescent lamps are generally the most efficient for an interior general lighting system where the large lighted area of the lamp can be effectively utilized to distribute light. At high mounting heights or where the lighting area shall be optically controlled more tightly, smaller, higher intensity sources generally are more efficient. Table 27 provides general guidelines as to lamp performance and application characteristics.

7.5.3 Lamps and Color. Light sources do not emit light equally at all wavelengths in the visible spectrum, and color distortion results. Therefore, the light source color spectrum should be carefully evaluated. If possible, an actual system or a demonstration area should be used to see how the colors appear before final installation. Two measures generally indicate the color characteristics of light sources: chromaticity and color rendering.

Chromaticity is an indication of *warmth* or *coolness* of a light source and is measured in terms of a color

Table 27
Lamp Output Characteristics

Interior General Lighting (Commercial and Industrial) Lamp Selection Guide

Lamp	Incandescent		Fluorescent				High-Intensity Discharge			
Type	Standard	Tungsten Halogen*	Cool White*	Warm White	Deluxe Cool White	Deluxe Warm White	Deluxe White Mercury	Warm Deluxe White Mercury	Metal Halide	High-Pressure Sodium
Application	Local lighting accent, display. Low initial cost general lighting	Accent, display general lighting in lobbies, theaters, etc	Most commercial and industrial general lighting systems	Many commercial and industrial general lighting systems	General and display lighting where good color daylight-like atmosphere important	General and display lighting where good color incandescent-like atmosphere important	Store and other commercial and industrial—general lighting	Store and other commercial—general lighting	Industrial, commercial general lighting. pools, arenas	Industrial general lighting. Some commercial applications
Lamp optical controllability potential	Excellent	Excellent	Fair	Fair	Fair	Fair	Good	Good	Excellent	Excellent
Glare control with typical luminaire	Good	Excellent	Good	Good	Good	Good	Excellent	Excellent	Good	Good
Color effects: accents	Warm colors	Warm colors	Yellow, blue, green	Yellow, green	None-excellent color balance	Yellow, orange, red-like incandescent	Red, yellow, blue, green	Red, orange, yellow, green	Yellow, green, blue	Yellow, orange
Color effects: grays	Blues	Blues	Reds	Red, blues	None	Blues	Deep reds	Deep reds, blues	Reds	Deep reds, greens, blues
Range of lamp wattages for typical applications	30-1000	100-1500	20-215	20-215	20-215	20-110	40-1000	175-1000	175-1500	35-1000
Range of initial lamp lumens for typical applications	210-23 740	3450-34 730	1300-16 000	1300-15 000	850-11 000	820-6550	1575-63 000	6500-58 000	14 000-155 000	3600-140 000
Initial lamp lumens/watt	7-24	17-24	65-85	65-85	40-60	40-60	30-65	35-60	80-100	100-140
Average rated lamp life (h)	750-2000	1500-4000	9000 (3 hours/start) — 30 000 (continuous burning for F40 types)				16 000-24 000+	24000+	75 000-20 000 (10 or more hours/start)	16 000-24 000+ (10 or more hours/start)
Light output depreciation characteristics	Good	Excellent	Good	Good	Fair	Fair	Fair	Fair	Fair	Excellent

*Light white is a newer more efficient substitute for cool white in many commercial and industrial applications. It has the same chromaticity but slightly poorer color rendering characteristics.

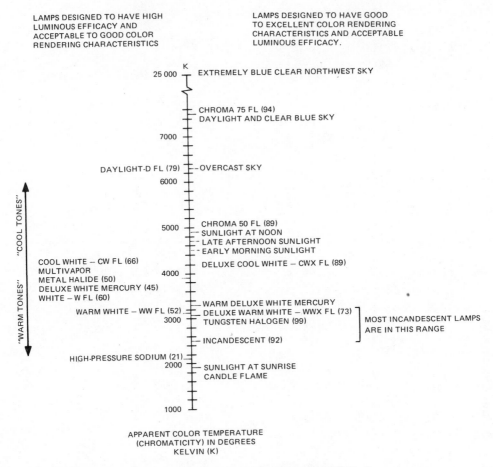

Fig 22
Color Characteristics of Light Sources

temperature scale in kelvin (see Fig 22). The higher the apparent color temperature of a light source, the cooler it appears to the eye. Sources with high color temperatures seem to be preferred at high illumination levels but personal preference is also involved.

Color rendering describes how well the light source makes object colors appear according to a defined standard. Often this is highly subjective and people refer to their own experience or to outdoor day-light conditions as their personal standard. Light source spectral distribution characteristics provide a clue to the color rendering ability of light sources. If the visible spectrum is generally filled between the limits of human vision, wavelengths from 400 nm to approximately 750 nm, color rendering will be better than if the source has emissions only in a narrow band or portions of the visible spectrum. Side by side visual comparisons and test installations are, however, a more reliable guide to color rendering and color acceptability. The CIE color rendering index (Ra) is a frequently used measure of a light source's ability to render colors in a natural and normal way. It may also be used to compare light source colors if the sources have the same chromaticity and meet certain other criteria.

7.5.4 Light Sources

7.5.4.1 Incandescent. Incandescent lamps have many desirable characteristics; however, they are very inefficient converters of electricity into light. Approximately 90%-95% of the energy consumed by an incandescent light source is converted not to light, but to heat. The efficacy of incandescent (general lighting types) is approximately 20 lm/W.

Recent technical advances have made possible slightly more efficient reduced wattage incandescent lamps. Improved mount structures, smoother tungsten filament wire, and the use of fill gas such as Krypton work together to provide the same lumen and life characteristics as previous lamps, but with approximately 5% less wattage.

Energy-saving potential also exists for incandescent lamps when reflector lamps can be employed. Incandescent lamps using built-in reflectors offer better utilization of the light produced compared to nonreflector types. In this family of lamps are the R lamps or indoor reflector lamps, PAR, or parabolic aluminized reflector lamps, and a newer line of indoor reflector lamps called ER or elliptical reflector types. ER lamps permit lamp wattage reductions up to 50% by improving reflector's optical efficiency.

7.5.4.2 Fluorescent. Fluorescent lamps, introduced in the late 1930s, offered the user two major benefits compared to incandescent lamps: increased efficiency and significantly longer life. Typical standard fluorescent lamps today have efficacies over 80 lm/W. The characteristics of modern fluorescent lamps are

High lumen per watt (LPW) efficacy (80+ LPW)

Long life (20 000+ hours)

Availability in a wide variety of sizes and shapes

Rapid starting and relighting after momentary outage

Availability of high color rendering types for color critical applications

Good maintenance of light output over lamp life.

Since the early 1970s, there has been a line of reduced wattage replacements for standard fluorescent lamps. These lamps are now available in all popular sizes and colors for most applications. However, limitations of the reduced wattage lamps are

Use only where ambient temperature does not drop below 60 °F (16 °C)

Use only on high-power factor fluorescent ballasts (CBM or other approved types)

Should not be used where cold air will be directed onto the lamp surface.

Wattage reductions possible with these reduced wattage lamp types are in the range of 10%-20% for the popular types. Light output is also reduced by 3%-18%, depending upon the specific lamp used. Typical energy savings will be approximately 5 W to 6 W per lamp for the popular 4 ft 40 W replacement and 17.5 W per lamp for the popular 8 ft slimline 75 W replacement. Savings in energy costs normally pay back the new lamp cost in one year.

Another type of fluorescent lamp available for further wattage and light output reductions is the impedance modified fluorescent. These lamps contain capacitors which add impedance to the lamp circuit and thereby reduce the current and wattage of the lamps. These lamps operate like a *fixed-dimmer* control and allow fluorescent lighting systems to be reduced in wattage by 33%, 50%, or more. Light output is also reduced in proportion. These special lighting devices may be considered where the lower light levels can be tolerated or where energy guidelines are mandated by legislation.

Growing in popularity are fixtures with three lamps in the standard 2 by 4 ft fixture. For maximum efficiency the fixture uses low-loss ballasts, cone-on-cone diffusers, and special reflectors. Switching provides further potential energy reductions. A second circuit permits the inside, the two outside, or all three lamps to operate — giving fixed 33%, 66%, and 100% light output adjustment.

Application limitations of fluorescent lamps, including energy-saving types are:

(1) Limited, normally, to indoor applications due to large optical size of the lamps

(2) Temperature sensitive: high and low ambient temperature reduce light output

(3) Highest wattage lamp: 215 W

7.5.4.3 High-Intensity Discharge.

High-intensity discharge or HID light sources fall into four major categories — mercury, metal halide, high-pressure sodium and low-pressure sodium. The HID group of sources offers many energy-saving opportunities due to high lumen per watt potential.

Mercury light sources have been basically available in their present form since about 1934. The efficacy and life have steadily increased to a point where they now have efficacies up to 63 lm/W and typical average life ratings exceeding 24 000 h. Improvements in phosphor technology have made mercury lamps available with color rendering properties suitable for all but the most critical applications.

Mercury lamps are available in numerous wattages ranging from 40 W to 1000 W in several phosphor colors. Typical applications include street lighting, industrial lighting, area and security lighting, and merchandising lighting, both indoors and outdoors.

The newer HID lamps, however, offer significantly higher efficacy and even better color rendering properties than mercury types now available. Thus, economically, mercury has been surpassed, and few new mercury systems are being installed where cost and energy efficiency have been considered.

Metal-halide lamps were introduced in the mid-1960s and have continued to gain wide acceptance for a broad range of applications, industrial and commercial, indoor and outdoor, and have been almost universally accepted for sports lighting applications.

While operating in a similar manner to mercury lamps, metal-halide types utilize rare earth metal iodides or halides to improve both efficacy and color rendition. Metal-halide lamps offer efficacies today up to 125 lm/W of *white* light. The color rendering of these sources allows their use in virtually all applications where high light output and good color are required. The broad color spectrum of metal-halide lamps coupled with the excellent optical control potential of the sources, makes them ideal for floodlighting, area lighting, and high-quality industrial lighting.

Limitations of metal-halide lamps include relatively long restart or restrike time (8 min–15 min) if the lamp is turned off momentarily. Another limitation is that low-wattage lamps are not presently available. Typical wattages are 175, 250, 400, 1000 and 1500. Wattages under 100 would increase the application potential for low mounting height.

Typical rated lamp life is 7500 h to 20 000 h with fair to good maintenance of light output over its life.

High-pressure sodium (HPS) lamps offer efficacies which exceed those available from most metal-halide lamps. Current available types offer efficacies up to 140 lm/W. A broad range of wattages is available with types now ranging from 35 W to 1000 W.

The main advantage of HPS compared to mercury and metal halide is shorter hot restart time. Lamps typically restart in 2 min–3 min after a momentary outage (see Table 28).

Major disadvantages of high-pressure sodium lamps are

(1) Color rendering index is low (typically 25), limiting applications for this source. This can be improved to a CRI of approximately 55-60, but at the expense of reduced life and efficacy.

(2) Lamps should be removed from circuit when lamps begin to cycle ON and OFF due either to premature or normal end-of-life failure. This is a lamp/ballast compatibility problem which will eventually be resolved by the ballast manufacturers.

(3) Color temperature is relatively low at 1900 K–2100 K

(4) Stroboscopic effects are greater than other HID lamps.

Typical applications of high-pressure sodium lamps include lighting for

(1) Streets and highways

(2) Industrial interior

(3) Outdoor area (parking lots, storage, sports, and recreational)

(4) Bridges and tunnels

(5) Closed-circuit television

HPS is also now being used for many office and commercial interior applications, sometimes coupled with metal-halide sources. A combined HPS/metal-halide system offers high efficacy and substantially improved color rendition. The life rating of typical HPS lamps is 24 000[+] h for most types.

7.5.4.4 Low-Pressure Sodium. Low-pressure sodium (LPS) is, at this time, the most efficient light source with lamp efficacies ranging up to 183 lm/W. Sizes available are 18 W, 35 W, 55 W, 90 W, 135 W, and 180 W. Major applications of LPS are security, area, and roadway lighting.

The low-pressure sodium (LPS), is more closely related to fluorescent than HID since it is a low pressure, low-intensity discharge source and is a linear source like fluorescent. As with fluorescent, this source does not lend itself to applications where a high degree of optical control is necessary such as *long throw* floodlights.

Table 28
Lamp Start Times

Lamp	Initial Start*	Restrike*
	(min)	(min)
Mercury	5→7	3→6
Metal Halide	3→5	10→15
High-Pressure Sodium	3→4	1

*Time to reach 80% lumen output.

The major disadvantage of low-pressure sodium is that it is a monochromatic source, that is, it provides light of only one color (yellow), which gives the lamp a poor color rendering index. Other disadvantages are high-ballast losses and low system efficiency. Major advantages are that this source delivers its full rated output throughout life and will restrike instantly when hot. Life ratings of low-pressure sodium sources are generally 18 000 h for all but the 18 W, which is currently rated at 10 000 h.

7.5.4.5 Economic Considerations. Although prices vary considerably among energy efficient light sources such as fluorescent and HID lamps, the major factor which determines lighting costs is not lamp or fixture cost but energy cost. Today this portion of the total cost of providing light with a general lighting system is typically 80% or more of the total. Before any decision is made concerning the choice of lighting system, the user should first decide what his color requirements are for the specific application. Once this is determined, he should then complete an economic analysis for the various light sources and systems he wishes to consider. The economic analysis should include the initial ownership and operating costs.

Most lamp and luminaire manufacturers offer free computerized analyses for this purpose and these sources of information should be consulted. In addition, many manufacturers offer a wide variety of literature on product applications and energy conservation for the prospective buyer. These sources will help the user make better economic judgments for specific applications.

7.6 Ballasts
    7.6.1 Figures of Merit. Electrical losses in ballasts may be calculated by determining the difference between the input power and the wattage delivered to the lamp.

$$\text{Ballast losses} = \left(\frac{\text{ballast input}}{\text{watts}}\right) - \left(\frac{\text{lamp}}{\text{watts}}\right)$$

Such information may be obtained from catalogs of ballast manufacturers. Note carefully the listings of the differences in ballast losses as a function of supply voltage. Ballasts designed for high-wattage lamps, which have higher starting voltages, may be more efficient on high-voltage distribution systems and vice versa. Carefully calculate the wiring and distribution losses and compare these losses against ballast losses to obtain the most efficient overall electrical system. Table 29 provides guideline information for common types of fluorescent and high-intensity discharge lamp ballasts.

7.6.2 Ballast Factor. The ratio between the light output from a lamp operating on a commercial ballast and the light output from the same lamp operating on a reference ballast. The ballast factor of a reference ballast is 100.

7.6.3 Fluorescent Ballasts
    7.6.3.1 Effect of Temperature and Voltage Variations. The lamp bulb wall temperature affects the efficiency and output of a fluorescent light more than any other environmental factor. The optimum temperature is 40 °C (104 °F). ANSI test procedures call for manufacturers to rate lamps in free air at an ambient temperature of 25 ± 0.5 °C (77 °F). To determine the light output of lamps in practical applications, tests certified by the Certified Ballast Manufacturers (CBM) organization may be employed. The ballast is operated at its rated center voltage and the initial lamp

Table 29(a)
Typical HID Lamp Ballast Input Watts

Lamp Type	ANSI Designation	Watts	Reactor	High Reactance Autotransformer (LAG)	Ballast Type Constant Wattage Autotransformer (CWA)	Constant Wattage Regulated (CW)	High Reactance Regulated (Regulated Lag)
Mercury	H46	50	68	74	74	—	—
	H43	75	94	91–94	93–99	—	—
	H38/44	100	115–125	117–127	118–125	127	—
	H39	175	192–200	200–208	200–210	210	—
	H37	250	272–285	277–286	285–300	292–295	—
	H33	400	430–439	430–484	450–454	460–465	—
	H36	1000	1050–1070	—	1050–1082	1085–1102	—
Metal-halide	M57	175	—	—	210	—	—
	M58	250	—	—	292–300	—	—
	M59	400	—	—	455–465	—	—
	M47	1000	1050	—	1070–1100	—	—
	M48	1500	—	—	1610–1630	—	—
High-pressure sodium	S76	35	43	68	—	—	—
	S68	50	60–64	88–95	95	—	105
	S62	70	82	127–135	138	—	144
	S54	100	115–117	188–200	190	—	190–204
	S55	150 (55 V)	170				
	S56	150 (100 V)	170	188	188	—	—
	S66	200	220–230	—	245–248	—	254
	S50	250	275–283	296–305	300–307	—	310–315
	S67	310	335–345	—	365	—	378–380
	S51	400	463–440	464–470	465–480	—	480–485
	S52	1000	1060–1065	—	1090–1106	—	—

Table 29(b)
Typical Fluorescent Lamp Ballast Input Watts

Lamp Type	Nominal Lamp Current	Nominal Lamp (W)	System Input (W)				Circuit Type
			Standard Ballasts		Energy-Saving Ballasts		
			One-Lamp	Two-Lamp	One-Lamp	Two-Lamp	
F20T12	0.380	20	32	53	—	—	Rapid start, preheat lamp
F30T12	0.430	30	46	81	—	—	Rapid start
F30T12, ES	0.460	25	42	73	—	—	Rapid start
F32T8	0.265	32	—	—	37	71	Rapid start
F40T12	0.430	40	57	96	50	86	Rapid start
F40T12, ES	0.460	34/35	50	82	43	72	Rapid start
F48T12	0.425	40	61	102	—	—	Instant start
F96T12	0.425	75	100	173	—	158	Instant start
F96T12, ES	0.455	60	83	138	—	123	Instant start
F48T12, –800 ma	0.800	60	85	145	—	—	Rapid start
F96T12, –800 ma	0.800	110	140	257	—	237	Rapid start
F96T12, – ES, 800 ma	0.840	95	125	227	—	207	Rapid start
F48 –1500 ma	1.500	115	134	242	—	—	Rapid start
F96 –1500 ma	1.500	215	230	450	—	—	Rapid start

4/22/83 RWW

RELATIVE LIGHT OUTPUT
VERSUS AMBIENT TEMPERATURE

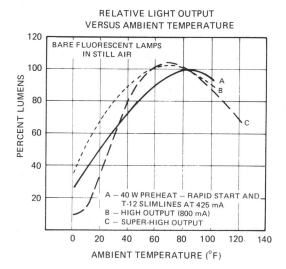

Fig 23
Relationship of Lumens
Versus Ambient Temperature

light output cannot be less than 95% of the manufacturers' rated output.

Rated lumen values are based on measurements made at an ambient temperature of 25 °C in still air. The effect of higher or lower temperatures vary with lamp type. Figure 23 shows the relationship of lumen variations versus ambient temperature for three types of fluorescent lamps.

Ballasts are typically designed to operate fluorescent lamps over a range of ±10% about the rated center voltage. The power input and light output tend to decrease with decreasing input voltage.

7.6.3.2 Energy Efficient Ballasts — General Concepts. In general, the greater the ballast size (power rating) the greater is the ballast efficiency. That is, the relative ballast losses are less for a 40 W ballast than a 20 W ballast. A two-lamp bal-

last is more efficient than a one-lamp ballast. The ballast efficiencies can generally be calculated from the manufacturers' catalog data. However, the designer shall be certain the manufacturer specifies the test conditions at which the ballast is rated.

The internal ballast losses are determined by coil construction, the nature of the magnetic materials, and resistances of the conducting coil wire. Over the years, standard ballast designs have been the victims of cost reduction which have tended to decrease ballast efficiencies. With increasing energy costs, ballast manufacturers have introduced energy efficient ballasts that minimize the ballast losses.

The following sections detail some of the new ballasts.

7.6.3.3 Low-Energy Ballasts. Fluorescent lamps may be operated at less than rated power and output, provided starting voltage and operating voltage requirements are met. In the case of rapid start lamps, and cathodes shall be heated regardless of the current through the lamp to provide rated lamp life. *Low energy* ballasts are low-current designs which can provide energy reductions compared to the standard units. They are useful where certain luminaire spacing to mounting height criteria shall be followed, and the desired illumination level is less than that obtained by full output operation of the lamp. Before operating fluorescent lamps at less than the rated output, it is wise to check the ambient temperatures of the air surrounding the lamps. Higher minimum ambients are required when low energy lamps or standard lamps are used with low-energy ballasts.

7.6.3.4 High/Low Ballasts. This type of ballast, which is generally available only for rapid start circuit operation, contains extra leads which can be connected or switched to provide multi-level operation of the lamp. Two-level and three-level rapid start ballasts are available, and fixture output may be set according to the lighting requirements of the area. Operation of fluorescent lamps at less than rated output may raise minimum operating temperature requirements.

7.6.3.5 Low-Loss Ballasts. Ballasts may be designed to reduce internal losses by improving mechanical and electrical characteristics. More efficient magnetic circuits, closer spacing of coils, and improved insulation systems can result in loss reductions of approximately 50%, compared to conventional units. Again, the ballast manufacturers' ratings should be consulted to determine which ballast will have the least losses for the power system and lamp combination involved.

7.6.3.6 Electronic Ballasts. The efficacy of fluorescent and HID lamp systems can be improved even more by utilizing electronic ballasts. Improvements occur in two ways

(1) By way of lower internal losses within the ballasts

(2) By making use of the fact that fluorescent lamp efficacy increases as a function of the frequency of the applied power. Electronic ballasts usually operate lamps at frequencies above 20 kHz, so the net result is approximately 10% improvement in lamp efficacy.

7.6.4 High-Intensity Discharge Ballasts

7.6.4.1 General. Ballasts factors for HID lamps are usually close to unity, and ballast circuitry is somewhat simpler than in fluorescent ballasts, due to the widespread use of single-lamp circuits.

For mercury vapor lamps, ballast circuits may be of the reactor, autotrans-

former, or regulator types. Regulator ballasts contain circuitry which is designed to operate the lamp at a relatively constant wattage even though nominal input line voltage may vary.

As a rule of thumb, HID ballast losses range between 10% and 20% of lamp wattage. However, since each ballast circuit is somewhat different, catalog ratings should be used for more precise information.

7.6.4.2 Ballasts for High-Pressure Sodium Lamps. At present, four general types of ballasts are available to the HPS lamp user. Each has its advantages and disadvantages.

The lag circuit ballast is simply an inductance placed in series with the lamp. It has the poorest wattage regulation due to changes in line voltages but can be used effectively on circuits where the line voltage varies by no more than ± 5%. It has a relatively high starting current which can produce a desirably faster lamp warm-up. It is relatively inexpensive and small in size and has low power losses.

The lead circuit ballast is built with a capacitance in series with the inductance and the lamp, in combination with an autotransformer. The ballast size is kept small by keeping a portion of the primary winding common to the secondary winding. With the secondary winding providing the wattage regulation, the effectiveness of the regulation depends on the amount of coupling between the primary and secondary. A variation of line voltage by ± 10% will only result in a ± 5% variation in lamp power. The lead circuit ballast has low starting current.

The magnetic-regulator ballast is essentially a voltage regulating isolation transformer with its primary and secondary windings mounted on the same core,

TYPICAL REGULATION CHARACTERISTICS

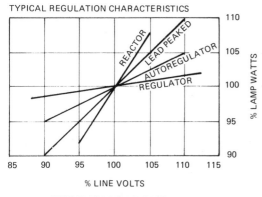

% LINE VOLTS

EFFECT OF LINE VOLTAGE
VARIATION ON LAMP WATTS
FOR VARIOUS BALLAST TYPES

Fig 24
Ballast Regulation Characteristics

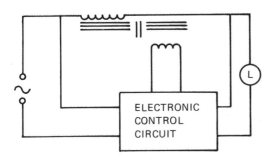

Fig 25
Energy Efficient Electronic
Controlled Ballast

and contains a third capacitive winding which adjusts the magnetic flux with change in either primary or secondary voltage. It provides the best wattage regulation with change of either input voltage or lamp voltage. However, it is the most costly and has the greatest wattage loss. Figure 24 shows the effect of line-voltage variation on lamp watts for various ballast types.

The major problem with existing HPS lamp ballasts is their inability to operate the lamp at rated power. Electronically controlled ballasts have been designed and built with a solid-state control circuit and a reactor. Figure 25 shows the circuit of one such ballast called the *super-regulated* ballast. The use of a solid-state switching device permits the control winding to be shorted in a *phase controlled* manner, and thus provides a smooth and continuous variation in the average inductance of the ballast. The solid-state control circuit monitors lamp and line operating conditions and then establishes the proper value of ballast inductance required to operate the lamp at its rated power. If the ballast operates the lamp so that it averages reasonably close to its rated wattage or below over its life, satisfactory performance can be expected.

Because the HPS lamp requires more voltage as it ages, special attention shall be directed to the regulation of the lamp wattage, which can vary to limits that can be very damaging to the lamp. To avoid this, lamp-wattage limits have been imposed on ballasts by the lamp manufacturers. The wattage and voltage limits are defined by the generally accepted diagram called the *trapezoid*. The trapezoid shows the lamp user of the range of wattage and voltage at which the lamp shall operate to give acceptable performance in life, luminous output, and stability. The trapezoid for the 400 W HPS lamp is shown in Fig 26. For present-day ballasts operating at rated input voltage, all lamps of a particular design will operate on a smooth curve within the trapezoid called the ballast characteristic. A properly designed ballast will generate a smooth wattage versus voltage curve with a *haystack* appearance so that with increasing voltage the watt-

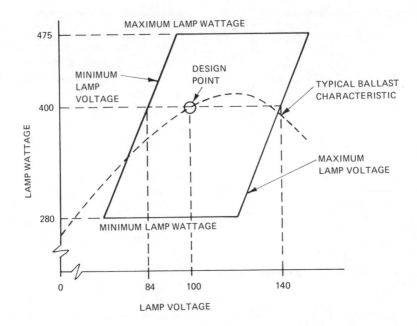

Fig 26
Trapezoid Diagram for the
400 W HPS Lamp

age will gradually rise to a peak, then start to decrease toward the end of life. HPS lamps on electronic ballasts can be made to traverse the trapezoid by way of a straight line. If the light output of such a combination is constant, input watts can be reduced approximately 20% over the life of the lamp.

7.6.4.3 Ballast Interchangeability. The optimum performance of any HID lamp can be obtained only by operating it with a ballast designed for the lamp. However, in recent years a number of lamps have been developed which can be used on various types of existing ballast circuits. Special metal-halide lamps are available which are designed to be used with mercury ballasts; a special type of high-pressure sodium lamp called *Penning Start* can also be operated on mercury ballasts.

It may be desirable to upgrade the efficiency of a lighting system by replacing a mercury lamp with either a metal-halide or a high-pressure sodium type which operates at higher lumens per watt. The light generated is, therefore, available at lower cost and it may be possible to reduce the number of luminaires depending upon mounting height, spacing, and fixture considerations. Some retrofit lamp designs, however, are not as efficient as their standard counterparts. They may provide somewhat less luminous efficacy, and may have shorter life. Lamp ratings should be carefully checked and any limitations on ballast use noted.

7.6.5 Ballast Life. Fluorescent and HID ballasts are long lived and may be expected to last 12 to 15 years in normal operation, provided nameplate ratings are not exceeded. However, high temperatures are detrimental to ballast life and a 10 °C increase in the hotspot temperature of a fluorescent lamp ballast may reduce expected life by 50%. Retrofitting old low-efficiency ballasts with new high-efficiency units may provide a double bonus: energy savings and longer ballast operating life since more efficient ballasts generally operate at lower temperatures.

Depending upon the circuit used, it may be important to ballast life to promptly replace failed lamps. This is particularly true of instant start fluorescent systems where failed lamps can *rectify*, drawing heavy current through ballast windings which will result in ballast failures. If lamps are removed from sockets to reduce lighting energy use, ballasts will normally remain energized except in the case of slimline or instant start circuits where removing the lamp disconnects the ballast from the electrical system. For other fluorescent and HID systems, an energized ballast will draw some power and may negatively affect system power factor. Rapid start fluorescent ballasts should be de-energized if lamps are removed to prevent ballast failure.

7.7 Luminaires

7.7.1 Efficiency Criteria. The principal purpose of a lighting fixture or luminaire is to contain the light source with its necessary mounting and electrical accessories and then direct the light from the source into the area to be lighted. Luminaire efficiency is defined as the amount of light that is emitted from the luminaire divided by the amount of light generated by lamps. Efficiency, therefore, tells how well the luminaire succeeds at permitting the light generated by the lamps to escape absorp-

tion, internal reflections, and trappings by the lamps themselves and other luminaire components.

7.7.2 Glare Control and Utilization. The most efficient luminaire may be no luminaire at all since exposed lamps may efficiently direct their light into the space without utilizing any type of enclosure. Such lighting equipment, however, can result in excessive glare and, depending upon the room reflectances, shape of the space, and mounting height, may not put the light where it is required. There is a figure of merit for general lighting systems which tells how effectively the luminaire and the room work in combination. This is the *coefficient of utilization* (CU) and is a measure of the light flux reaching the task area divided by the light flux generated by the lamps. The coefficient of utilization depends upon the intensity distribution of the luminaire, size and shape of the space, reflectances, and luminaire mounting arrangement. CU data are normally provided by the luminaire's manufacturer for each type of luminaire. The CU permits rapid estimation of the illumination that can be achieved for a given system.

Where people shall spend long periods of time doing visual work, it is important to control the glare from the lighting system. The *visual comfort probability* or VCP system is a measure of direct discomfort glare based upon the studied reactions of people to areas of brightness within their view. A VCP rating indexes the probability that a lighting system will be comfortable from the direct glare standpoint if a person stands in the worst position in the room and looks horizontally. VCPs can also be calculated for individual positions and viewing directions within a space.

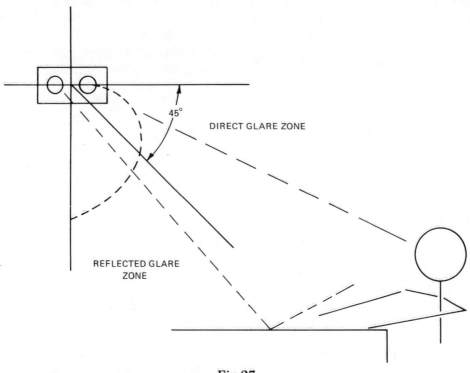

Fig 27
Glare Illustration

Generally, light emitted from a luminaire within a cone described by an angle of 45° rotated around a vertical line from the luminaire to the floor will not contribute substantially to direct glare. Light emitted above 45° may cause glare depending upon its intensity and the luminance of the surrounding space. Figure 27 shows that luminaires which emit light below 45° can cause loss of contrast on the tasks, depending upon their position in the field of view. To minimize veiling reflections, light emitted from luminaires above 45° is desirable. Thus, a compromise shall be made to avoid having light from the luminaire emitted between 0° and 45° to minimize veiling reflections and between 45° and 90° to minimize direct glare. The type of task or use of

the space determines which zone should receive the greater emphasis.

7.7.3 Shielding Media. Some unique solutions to the problem described above have been provided using new types of shielding materials which can direct light from luminaires to rather narrow zones for maximum visual effectiveness and control of glare. One example of this is a lighting material with a so-called "batwing" distribution. Figure 28 shows that a high percentage of light is emitted near 45° so that if work stations are placed between rows of luminaires the major illumination on the task comes from the side. Veiling reflections will thus be minimized with a fair degree of glare control.

Other techniques can also minimize veiling reflections. Polarizing materials

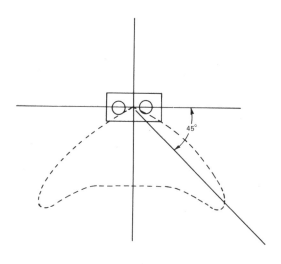

**Fig 28
Batwing Distribution**

can reduce the amount of light reflected from the task, thus improving task contrast. Changing the position of the task with respect to the luminaire can result in dramatic improvement, and if task locations are known, installing luminaires away from the reflected glare zone will always be beneficial.

7.7.4 Dirt Effect and Maintenance Considerations. The efficiency of the luminaire is highly dependent upon the reflectance of its interior and exterior surfaces and the amount of dirt collected on the lamps and shielding materials. All surfaces eventually collect dust, or otherwise lose their lighting effectiveness over a period of time. In extreme cases, such losses may amount to as much as 50% over a two-year period. More typical, perhaps, are losses of 10%-25% over the same period in air-conditioned spaces 8 h-12 h/day usage.

Luminaire manufacturers publish expected *luminaire dirt depreciation fac-tors* (LDD) to help evaluate loss of efficiency due to dirt. However, it shall be recognized that lighting in an interior space depends not only on luminaire dirt depreciation but also on room surface dirt depreciation, for example, aging and discoloration of luminaire shielding materials and permanent degradation of room and furniture surfaces.

7.7.5 Air Movement. Enclosed luminaires may "breathe," pushing air out of their enclosures as lamps warm up and drawing air in as lamps cool down. This may increase the dirt collection on the interior surface of the luminaire. In dirty areas dirt accumulation can be substantially lessened by using luminaires which circulate room air, using heat from the lamps to provide a chimney effect. This draws dirt up and through the luminaire without depositing it on luminaire surfaces. Air handling luminaires, that is, those luminaires which are connected to a building HVAC system, are usually constructed so that room air is exhausted by passing it over lamp, ballast, and luminaire surfaces. Not only does this lessen dirt accumulation, but it can make the lamps operate closer to their peak output. Ballasts are also kept cooler, which extends life.

7.8 Lighting Controls

7.8.1 General. The proper control of a lighting system is one of the most effective ways to save lighting energy. Control techniques may range from the simple ON/OFF switch installed on an individual luminaire to elaborate computer-controlled master switching systems designed for operating the lighting system of a large building. The installation of a proper number of switchpoints is generally the key to effective control of energy by switching. Once a system is installed, additional switchpoints may be difficult to add; so careful attention should be paid to this part of the lighting system during initial design.

7.8.2 Switching. It has been common practice to wire electrical power systems so that the minimum amount of electric wire or circuiting is required. With a very small increase in cost, electric circuits can be wired so that full advantage of reduced lighting can be obtained.

Switching should be arranged for each area in which an individual or group might work and require higher levels of lighting than the balance of the room. Low-voltage switching techniques increase the multiple point control ability of the system; 277 V switching with barriers between different switch legs is available for use in commercial buildings.

Partial lighting could be acceptable at several periods of the day: during cleaning operations, at lunch hour, and before the staff is fully occupied in the morning. For example, office lighting may range from 50 fc to 100 fc, whereas janitors can work perfectly well and safely at 15 fc to 20 fc. The control is normally arranged so that half of the ballasts or one-third of the ballasts in a fluorescent or HID installation are turned off. In some cases, this means that half, one-third, or two-thirds of the fixtures will be extinguished. In other cases where multiple ballasts are installed in continuous rows of fixtures or paired fixtures, this simply means the fixture will be dimmed to one-half, one-third, or two-thirds light. In the latter case, the advantage of even lighting and normal ratios of minimum to average lighting are obtained and minimum shadowing occurs. This type of control is very easily accomplished by wiring branch circuits on alternate switch or circuit breaker legs or in a three-phase system by wiring

one-third the lights on each phase. Suitable arrangements which stagger the lighting can provide a current balance at the electric service. In most cases, very little additional conduit is required, and only a minimum amount of additional wire and a few additional branch breakers for each panel are needed.

It may be desirable to place switches adjacent to windows so that lights can be turned off or reduced if not needed. Significant energy savings can be made by the use of day lighting. In some instances, two zones of control, running parallel to the windows, can be justified.

It is essential to remind those responsible for controlling the light switches of the need to turn off lights. No advantage will be gained in the introduction of additional controls if they are not utilized. The use of posters, notices, meetings, and checking by supervisory staff is desirable.

7.8.3 Dimming. Various methods of dimming are available which provide a pleasing effect, utilize only the amount of light needed, and save energy. Reduced voltage is mostly used for incandescent, but more sophisticated methods are required for fluorescent and HID lighting.

It is possible to reduce the voltage of a system very simply by the use of variable tap autotransformers or by delayed firing thyristors.

Because variable tap autotransformers have relatively low losses and may be used to control either large or small incandescent loads, such control can be used for theater or accent lighting where varying levels of illumination are desired. A less expensive way of accomplishing the same type of lighting control is to use fixed tap autotransformers; however, the lighting levels are adjustable only in steps.

A second type of incandescent dimming uses transistors or thyristors. They reduce the voltage by reducing the time in each cycle the voltage is applied. This method is the least expensive. One other form of voltage dimming involves the use of the saturable reactor together with an electronic control.

Dimming of the HID and fluorescent lamps can be accomplished with a phase controlled circuit. In full range systems using rapid start fluorescent lamps, lamp cathodes are held at relatively constant voltage to prevent premature failure of the tube or extinguishing of the arc. Some types of electronic ballast also incorporate a dimming feature with the additional advantage that power losses through the solid-state devices are minimal.

New types of auxiliary controls were developed in 1981 which can cut the power of lamps operating on standard ballasts by 25% or more. These controls are easily installed in the luminaire or branch circuit, but a careful evaluation should be carried out beforehand to make sure that the control, ballast, and lamp work effectively without compromising system performance or component life.

7.8.4 Remote Control Systems. In most large buildings it is desirable to have the ability to control lighting from a central location or from central floor points. This enables individuals assigned the task to turn on light in the morning and turn off light at night, and make intermittent adjustments to the lighting as desired.

7.8.5 Automatic Control System. Automatic control systems permit programmed operation of building lighting. The switching function can be activated by a time clock, photocell, presence detector, or a programmable controller.

7.9 Optimizing Lighting Energy. To obtain optimum use of energy in lighting systems, the most important considerations in the lighting-design process are

(1) The person and the visual task involved

(2) Quantity and quality of illumination

(3) Lighting hardware (lamp, ballast, luminaire, control)

(4) Maintenance characteristics and procedures

The relative importance of each of these factors varies with the situation. For example, when selecting lighting fixtures for a parking lot, the visual task, although undemanding, is all pervasive. That is, the entire parking lot requires a low uniform wash of light while the construction characteristics dictate as few obstructions (poles) to the pavement surface as possible. On the other hand, the lighting of a conventional office require a lighting level of sufficient magnitude to permit reading of second or third carbon copies at specific work stations while the remaining portion of the space may be lighted to a somewhat lower level. The construction characteristics of such a space are usually much less limiting as far as equipment locations are concerned.

7.9.1 People and the Visual Task. The designed illumination level should be appropriate for the visual task. Some judgement shall be exercised in establishing the design level because recommended standards may specify either a single value or a range. These values are starting points and should be adjusted according to the age of the person doing or likely to be doing the task, the visual difficulty of the task and its importance. Where a task is classed as "visually casual," but is in the vicinity of a more demanding one, care should be taken to ensure that maximum contrast ratios are not exceeded in the desire to keep illumination levels as low as practical.

7.9.2 Illumination Quantity and Quality. Quantity of illumination is usually measured in lumens per square foot (footcandles) or lumens per square meter (lux). Since the eye cannot "see" footcandles, some lighting recommendation are now given in footlamberts or units of luminance. Luminance is more closely related to seeing since the eye is an excellent judge of luminance differences, (contrast) a necessary element of vision.

Of equal importance is the quality of illumination. Quality factors include lighting uniformity, control of glare, direct and reflected, and proper color.

A common mistake is to compromise the quality of a lighting system to maximize quantity or efficacy (lumens/watt). This is counter productive since glaring, uncomfortable lighting results in complaints of headaches, eye strain, and similar problems which can affect the productivity.

7.9.3 Lighting Hardware. Energy efficiency is only one of the factors in the lighting hardware selection process. Physical characteristics, quality of construction, ease of servicing along with the lighting quality factors are most important in a given installation. However, particular attention should be given to the parts of the system which transform electric energy into visible energy (the lamp and ballast) and the luminaire which collects and directs that visible energy into the room. Compensation for an inefficient ballast, lamp, or luminaire cannot generally be provided by manipulation of the other variables in the lighting system.

7.9.4 Maintenance Characteristics. Fixture selection cannot be made without considering maintenance techniques. Lower initial lighting levels are permissible when the lighting system is properly maintained, because initial levels are maintained. Ignoring maintenance will require higher initial lumens, more fixtures or higher wattage lamps, and will be more expensive because of wasted energy. When costs are calculated in units of annual dollars per footcandle, an evaluation can be made to aid the selection.

7.9.5 Space-Mounting Height Ratio. The ratio of luminaire spacing to height of the luminaire above the work plane is a critical consideration in industrial lighting design. This relation is known as the spacing to mounting height ratio (S/MH). The value S/MH in lighting fixture manufacturer catalogs is the maximum spacing that will result in relatively uniform lighting at the work plane. This spacing will provide good overlap of light between adjacent fixtures, regardless of interference from workmen or equipment. The wider the fixture spacing, the greater shall be the light output from each fixture. Many industrial plants are congested with columns, equipment, and piping which cast shadows. For this reason, individual work stations shall be considered in addition to area lighting.

High-intensity discharge lamps are available in lumen output sizes that span a range of more than 10:1. Initial cost of a lighting system will be lower and energy costs will be less if larger lamps and fewer luminaires are used. However, caution is required because using larger lamps and spacing luminaires far apart can produce unsatisfactory light distribution. At a spacing to mounting height ratio of 1:1, for example, the overlap of light from adjacent fixtures and lamp shielding will be reasonable

Table 30
Energy Requirements for Four Major Lighting Systems

System*	Fluorescent	High-Pressure Sodium (400 W)	Metal Halide (400 W)	Mercury (400 W)	Incandescent (1000 W)
No fixtures	73	40	65	118	70
Power requirement	33.2 kW	19 kW	30 kW	52 kW	70 kW
% of HPS power required	175%	100%	158%	274%	368%

*System requirements are for an entire 10 000 ft^2 area. Numbers are total except the last row which is in percent.

and reflector brightness will be within acceptable limits. When the mounting height is equal to the spacing between fixtures, the area covered by each luminaire is equal to the square of the mounting height. Multiplying the square of the mounting height by twice the desired average maintained light level yields a good approximate value for required initial lamp lumens.

Table 30 offers a comparison of the energy requirements of four major lighting systems. The table summarizes the results of a fixture layout study to provide an even 100 fc to illuminate a 10 000 ft^2 area. The data would also hold true for similar interior lighting applications, provided there was sufficient ceiling height for proper mounting of the fixtures. Fluorescent lighting is the optimum choice in most low ceiling areas.

7.10 Power Factor. A high power factor is desirable for electrical power systems. For a given kilowatt load, the current required will be inversely proportional to the power factor. Minimum current has the following benefits:

(1) The ratings or ampacities of cables, switchgear, transformers, and related systems can be minimized

(2) The power losses in the cables and transformers which are proportional to the square of the line currents will be reduced, thus, saving energy

(3) Voltage drops which are proportional to the line currents will be reduced

Fortunately, lighting system elements either have inherently high power factors or are available with integral power-factor correction. Well designed lighting systems enhance the overall building power factor. Fluorescent ballasts used in commercial buildings deliver close to unity power factor. These ballasts use capacitors as an integral part of their construction. In fact, some ballast/lamp combinations exhibit a leading rather than a lagging power factor. Almost all ballasts are made with capacitors which not only provide high power factor but provide, in two-lamp units, phase shifting which reduces stroboscopic effects and in some circuits acts as a part of a constant wattage or voltage-regulating mechanism. While such high power-factor ballasts are almost routine for fluorescent lighting, for many of the HID luminaires they must be specified. In the absence of any specific negative criteria, it should be a general rule to specify high power-factor ballasts.

Incandescent lamps, almost purely resistive elements, have essentially unity power factor. Dimming equipment which is resistive in nature will not reduce system power factor; however, dimming equipment for discharge type lighting, including fluorescent and HID, may have an adverse effect. Simple voltage reduction equipment consisting of transformers and autotransformers will have negligible effect on the reactive power delivered to the system. Even when the lighting levels are low and the transformer exciting power (which is highly reactive) predominates, the overall reactive element of the system power is usually a small percentage of the system design power. Even in these systems, power-factor correction can be applied if studies show it to be desirable.

Most dimming systems for discharge type lighting depend on a wave-chopping action or use only part of the cycle to reduce power to the lighting system. These units will have an inherently lower power factor in the dimming phases because of their high harmonic content and relatively high component reactances. For these devices, power-factor correction capacitors will additionally reduce the high harmonic content of these systems, if properly designed. Improper use of power-factor correction can increase harmonics due to resonance.

7.11 Interaction of Lighting with Other Building Subsystems

7.11.1 General. The lighting subsystem in a building interacts primarily with both the electrical and HVAC subsystems. It has secondary effects on building acoustical controls and fire safety. Lighting reacts with the electrical system because it uses electricity as its source of energy, and it reacts with the heating, ventilating, and cooling subsystems because of the effects of the heat dissipated from the lighting subsystem. This heat generally reduces the building heating requirements.

7.11.2 HVAC Subsystem Interaction. Lighting for buildings is derived either from natural sources or from artificial electrically operated devices. Depending on building types and location, the balance between daylight and artificial light during daylight hours varies considerably. At night, artificial lighting is required in all applications.

Sunlight provides 36 lumens of light for each Btu per hour per square foot of solar radiation. By way of comparison, Table 31 shows this relationship for more typical interior lighting sources.

Thus, daylight provides light with less total associated heat than do most of the other commonly used interior lighting sources. One problem with daylight is that without proper control, more light can be introduced than may be required, and, during the cooling seasons at least, more heat is introduced than is needed or would be present with more limited artificial sources. This is particularly true in the direct beam of sunlit interiors.

There are very few studies detailing the effects of natural and artificial lighting on the energy required for heating and cooling buildings. One computer simulation analysis made for a high-rise office

Table 31
Comparative Output of Light Sources

Source	Lumens/(Btu/h) per Square Foot (including ballast)
Incandescent Lamp	6
Mercury Lamp	15
Fluorescent Lamp	20
Metal-Halide Lamp	30
High-Pressure Sodium Lamp	35
Daylight	36

building[15] shows that for each additional watt per square foot of peak lighting demand there is an increase of approximately $(0.6 \text{ kWh/year})/\text{ft}^2$ in cooling energy required; and a decrease for heating energy required of from $(0.5 \text{ kWh/year})/\text{ft}^2$ in cold climates to $(0.1 \text{ kWh/year})/\text{ft}^2$ in warmer climates. These results are based on a projected occupancy of 2500 equivalent full-load hours per year.

A computer simulation of this same prototypical office structure was made to determine the effects of daylight on the energy required for artificial lighting when using fluorescent troffers for weather conditions in St. Louis, Missouri. This study showed that compared with a building with venetian blinds closed to a shading coefficient of 0.6, and in which the perimeter lighting operates constantly during working hours, turning off perimeter lights and venetian blinds when appropriate, could save approximately 25% of the annual energy of the lighting system. Annual HVAC system cooling energy requirement decreases by 10% and the annual heating energy requirement increases by 7%. The overall net effect on the building, believed to be typical of many new high-rise office structures, by using daylight to the maximum feasible extent is to decrease the annual energy input to the building by approximately 7% (using 3414 Btu/kWh). It must be emphasized that these results pertain only to the building studied. The data could vary appreciably for other designs and other climate conditions.

---

[15] Energy Conservation Principles Applied to Office Lighting and Thermal Operations, *Federal Energy Administration Conservation Paper No 18*, Rev Dec 23, 1975.

7.12 Evaluation Techniques. In either new or existing installations there are usually numerous alternatives that should be examined to determine the optimum lighting system from the energy standpoint. Evaluation techniques which rank systems on the basis of watts/square foot, watthours/square foot, Btu/hour, etc, may all be applicable, given the great variety of design techniques, legal constraints, and policies which may apply. The economics of the lighting system must also be examined and as energy costs begin to more closely reflect the actual cost, scarcity, demand, and importance of various fuels, the life-cycle cost of a lighting system can be used as a good indicator of its efficiency. The heat from the lighting system should be included as a negative cost during heating months and the energy to remove this heat during cooling months added as an additional energy cost.

In the annual cost model (Fig 29), initial and operating costs are determined using actual or predicted values for hardware, installation, energy, and maintenance costs. An "owning cost" is added to the annual costs to annualize the initial cost of the installation. If the cost factors have been estimated accurately, the annual cost model provides a good estimate of actual money that will be spent.

7.13 Bibliography

[1] ANSI C82.1-1977 (R1982), American National Standard Specifications for Fluorescent Lamp Ballasts.

[2] ANSI C82.2-1983, American National Standard Methods of Measurement of Fluorescent Lamp Ballasts.

[3] ANSI C82.3-1983, American National Standard Specifications for Fluorescent Lamp Reference Ballasts.

Lighting System Parameter		Base	II
	1. Rated initial lamp lumens per luminaire		
	2. Rated lamp life (hours) at _____ hours per start		
	3. Group replacement interval (hours)		
	4. Average watts per lamp		
	5. Input watts per lumaire (including ballast losses)		
Basic Data	6. Coefficient of utilization		
	7. Ballast factory (fluorescent)		
	8. Lamp depreciation factor		
	9. Dirt depreciation factor		
	10. Effective maintained lumens per luminaire $(1 \cdot 6 \cdot 7 \cdot 8 \cdot 9)$		
	10A. Average footcandles on work surface $(10 \div ft^2/luminaire)$		
	11. Relative number of luminaires needed for equal maintained footcandles (10 of base system $\div$ 10 of system compared)		
	12. Net cost of one luminaire		
	13. Wiring and distribution system cost per luminaire		
	14. Installation labor cost per luminaire		
Initial Costs	15. Net initial lamp cost per luminaire		
	16. Total initial cost per luminaire $(12 + 13 + 14 + 15)$		
	17. Annual owning cost per luminaire (15% of $12 + 13 + 14$)		
	18. Relative initial cost for equal maintained footcandles $(16 \cdot 11$ of system compared $\div$ 16 of base system)		
	19. Burning hours per year		
	20. Number of lamps group replaced per year $(19 \cdot = $ lamps/unit $\div 3)$		
	21. Number of interim spot replacements $(20 \cdot = $ burn outs in GR interval)		
	21A. Number of lamps spot replaced per year — No group relamping $(19 \cdot $ lamps/unit $\div 2)$		
	22. Replacement lamp cost per year (20 or 21A $\cdot$ net lamp cost)		
Operating Costs	23. Labor cost for group replacements $(20 \cdot$ group labor rate/lamp) at $ _____ / lamp		
	24. Labor cost for spot replacements $(21 \cdot$ spot labor rate/lamp at $ _____ / lamp		
	25. Cost of cleaning per luminaire per year		
	26. Annual energy cost per year $(5 \cdot 19 \cdot$ ¢/kWH $\div$ 100 000 at _____ ¢ kWH		
	27. Total annual operating cost per luminaire $(22 + 23 + 24 + 25 + 26)$		
	28. Relative annual operating cost for equal maintained footcandles $(27 \cdot 11$ of system compared $\div$ 27 of base system)		
Total	29. Total annual cost — owning and operating — per luminaire $(17 + 27)$		
	30. Relative total annual cost for equal maintained footcandles $(29 \cdot 11$ of system compared $\div$ 29 of base system)		

Fig 29
Annual Cost Work Sheet

[4] ANSI C82.4-1978, American National Standard Specifications for High-Intensity-Discharge Lamp Ballasts (Multiple Supply Type).

[5] ANSI C82.5-1983, American National Standard for Lamp Reference Ballasts.

[6] ANSI C82.6-1980, American National Standard Methods of Measurement of High-Intensity-Discharge Lamp Ballasts.

[7] ANSI/ASHRAE/IES 90A-1980, *Energy Conservation in New Building Design* (Section 9).

[8] BURKHARDT, W. C. High Impact Acrylic for Lenses and Diffusers, *Lighting Design and Application*, vol 7, no 4, April 1977, pp 12-20.

[9] CHEN, K. and GUERDAN, E. R. Resource Benefits of Industrial Relighting Program, *IEEE Transactions on Industry Applications*, vol IA-15, no 3, May/June, 1979, pp 331-334.

[10] CHEN, K., UNGLERT, M. C., and MALAFA, R. L. Energy Saving Lighting for Industrial Applications, *IEEE Transactions on Industry Applications*, vol IA-14, no 3, May/June 1978, pp 242-246.

[11] ENERGY MANAGEMENT COMMITTEE of the Illuminating Engineering Society, *IEEE Procedure for Calculating Lighting Power Limits for New and Existing Buildings — (Unit Power Density Procedure EMS-6*, Illuminating Engineering Society, New York, July 1980.

[12] ENERGY MANAGEMENT COMMITTEE of the Illuminating Engineering Society, IES Recommended Procedure for Lighting Energy Management of Existing Buildings — EMS-4, *Lighting Design Applications*, vol 9, April 1980, p 30.

[13] JEWELL, J. E., SELKOWITZ, S., and VERDERBER, R. Solid-State Ballasts Prove to be Energy Savers, *Lighting Design and Applications*, vol 10, no 1, Jan 1980, pp 36-42.

[14] NUCKOLLS, J. L. Electrical Controls, *Lighting Design and Application*, vol 7, no 10, Oct 1977, pp 16-19.

[15] VERDERBER, R., SELKOWITZ, S. and BERMAN, S. Energy Efficiency and Performance of Solid State Ballasts, *Lighting Design and Application*, vol 9, no 4, April 1979, pp 23-28.

[16] WENGER, L. Architectural Dimming Controls, *Lighting Design and Application*, vol 7, no 6, June 1977, pp 19-21.

[17] IES Light Handbook, 6th ed, vols I and II, 1981.

8. Cogeneration

8.1 Introduction. Cogeneration may be described as an efficient method for the production of electric power in conjunction with process steam or heat which optimizes the energy supplied as fuel to maximize the energy produced for consumption.

In a conventional electric utility power plant, considerable energy is wasted in the form of heat rejection to the atmosphere through cooling towers, ponds, lakes, or rivers. In a cogeneration system, heat rejection can be minimized by systems which apply the otherwise wasted energy to process systems requiring energy in the form of steam or heat.

The energy-saving potential in cogeneration and the technical and economic benefits of the process are described in this section. Cogeneration is defined in this section as generalized groups of systems and methods by which electricity or shaft horsepower, or both, and steam are coincidentally produced in a more efficient manner than if each were produced separately and without consideration for the other.

8.2 Forms of Cogeneration. Cogeneration systems may be grouped broadly into two types generally referred to as *topping* cycles or *bottoming* cycles. Virtually all cogenerators use the *topping* cycle which generates electricity from high-pressure steam and uses the exhausted steam or other hot gas for process heat.

The *bottoming* cycle utilizes lower working temperatures in various arrangements to produce process steam or electricity. Thermal energy is first used for the process, then the exhaust energy is used to produce electricity at the bottom of the cycle. The applications for electrical generation may be limited, and this cycle is most attractive where large amounts of heat are utilized in processing, such as, in rotary kilns, furnaces, or incinerators.

Figure 30 illustrates, in a simplified

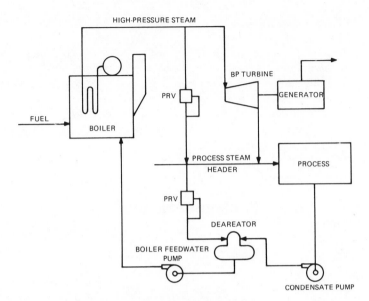

Fig 30
Plant Topping Cycle Cogeneration
Steam System

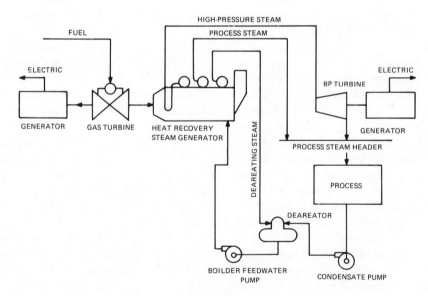

Fig 31
Plant Combined Cycle Cogeneration
Steam System

manner, a widely used topping cycle in many industrial process plants consisting of high-pressure boilers, typically 600 psig-1500 psig, generating steam for admission to back pressure steam turbines. The steam turbine either drives an electrical generator or serves as a mechanical driver for such equipment as fans, pumps, compressors, etc. The advantage of this system is that only the energy content of the steam required for mechanical power and losses is utilized in the turbine. The majority of the energy content remains in the back pressure steam which will be utilized in the process system. Electrical generation in the 4500 Btu/kWh-6500 Btu/kWh range is typical as compared to the usual 10 000 Btu/kWh-12 000 Btu/kWh heat rate of the electric utility. At 100% efficiency, the conversion rate is 3414 Btu/kWh. The use of this system implies a balance between kilowatt requirements and process steam requirements. If a balance does not exist, other means shall be provided to effect a balance or the difference shall be handled by outside means, such as exchanging power with the electric utility. The system illustrated in Fig 30 typically has an output of 30 kW-35 kW (1000 lb/h) of steam flow and has a thermal efficiency of approximately 80%.

Another highly efficient topping cycle employs the gas turbine-heat recovery boiler combination which may or may not utilize a steam turbine in the cycle. Occasionally, the exhaust from the gas turbine can be used directly in the process, as for certain lumber drying kilns.

The combined cycle steam, as shown in Fig 31, has a much higher kilowatt-producing capability per unit of steam produced than the back pressure system in Fig 30. Typically, the system illustrated in Fig 31 has an output of

300 kW-350 kW/(1000 lb/hr) of steam produced at a thermal efficiency of approximately 70%.

Advanced gas turbines available for base load service have exhaust temperatures in the 900 °F-1000 °F range with exhaust gas mass flows of typically 25 lb/kWh-35 lb/kWh resulting in some 5250 Btu/kWh of available exhaust energy. This high energy content can convert condensate to steam in a heat recovery boiler for use in a steam turbine or for direct use in process requirements. Generally, no additional fuel is required in the heat recovery steam generator.

The use of a steam turbine-generator results in additional incremental electrical power being generated at basically the cost of capital. There is some incremental reduction in the total quantity of waste heat steam produced because of the higher pressure level required. There is also an increase in the energy input to some small extent. In one system, the generation of approximately 7500 kW of incremental electrical power was possible at less than $250/kW installed cost (in 1976) with a before-tax return on investment of 50%.

The objective in cogeneration is accomplished by utilizing the heat rejection inherent in the cycles commonly used for the production of electricity or process steam. Figure 32 illustrates the temperature — entropy diagram which will be used to illustrate the basic cycle.

In both diagrams of Fig 32, Point A represents water conditons after the boiler feed pump, and A to B represents the energy addition in the boiler system. Point C represents the steam after going through the turbine or process and before being condensed. The area enclosed by ABCD represents the *work* portion of the cycle. The cross-hatched area under

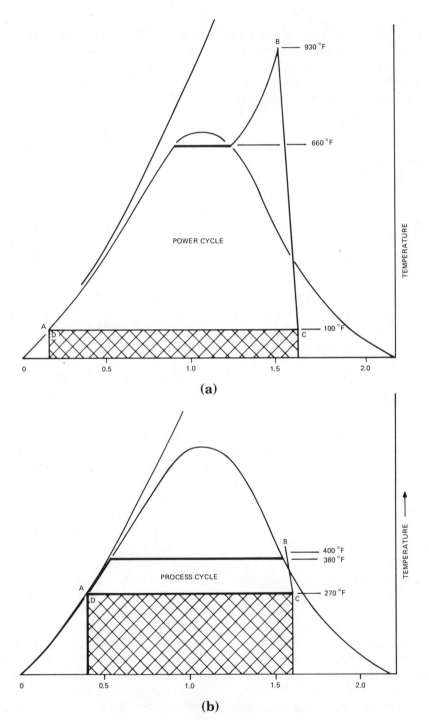

**Fig 32
Entrophy Diagrams for Generation
(a) Electric Output (b) Steam Output**

CD represents the rejected heat loss of the system.

The typical power cycle usually has a thermal efficiency in the 35% range. The condenser losses (heat rejection) are approximately 48%, and stack and miscellaneous boiler losses are approximately 17%.

The typical process steam generating cycle may operate at efficiencies in the 85% range with stack and miscellaneous boiler losses of approximately 15%.

The cogeneration approach attempts to minimize these heat rejection losses by combining the production of electricity and steam into a common facility. In a sense, the process load replaces the condenser so that useful energy is extracted from the exhaust steam. The overall efficiency is approximately 70%. However, it should be noted that each potential cogeneration facility will have unique requirements in the amount of electricity and steam required. Each facility will have varying degrees of energy savings.

8.3 Determining the Feasibility of Cogeneration. Eligibility requirements for cogeneration are based on economic and process requirements which will vary depending on the system's operating parameters. Major items for consideration are:

(1) Purchased fuel cost
(2) Purchased electrical cost
(3) Hours of operation per year
(4) Plant system size (kW)
(5) Amount of steam required (lb/h)
(6) Steam pressure level (psig)

Figure 33 illustrates an approximate approach to determine the feasibility of a given system. When all, or many, of the industrial plant factors are in the upper range, the feasibility of cogeneration appears good. If all, or several, of the factors are in the lower ranges, cogeneration may not be economically feasible. Note that the two columns, fuel cost and electric cost, should be compared, that is, if electric costs are 30 mil/kWh and fuel costs at the industrial plant are $2.00/MBtu, then eco-

Fig 33
Approximations for Determining
Cogeneration Feasibility

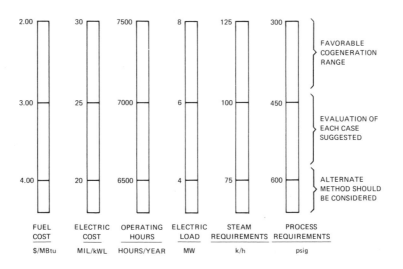

Table 32
Scale Cost Comparisons

Fuel Transportation

Annual Requirements	Cost Per Ton-Mile	Cost Per Ton*	Transport Cost Per K-lb/h Steam
0.1 M ton/y (carload shipments)	$0.024/ton-mile (1000 cars/y)	$24	$1.50
1 M ton/y (unit train)	$0.012/ton-mile (10 000 cars/y)	$12	$0.75
10 M ton/y (unit train)	$0.010/ton-mile (100 000 cars/y)	$10	$0.63

*For 1000 mile shipments, 100 ton/car, 10 000 Btu/lb coal. Cost of coal not included.

Coal Terminal Facilities

Terminal Size	Capital Cost Per Ton of Throughput		Fixed Cost Per K lb/h Steam*
(Approx boiler lb/h)			
0.1–0.5 M ton/y (0.2 − 1.0 M)	$10.00	$1–5 M	$0.14
1.25 M ton/y (2 500 000)	$ 9.60	$12 M	$0.13
5 M ton/y (10 000 000)	$ 4.00	$20 M	$0.05

*At 22% fixed charge rate, 8000 h/y, 1250 Btu input/lb steam.

Boiler Plant

Boiler Size	Boiler/Aux Cost	Installed Cost	Fixed Cost Per K lb/h Steam
100 000 lb/h	$ 1.5 M	$ 2.5 M	$0.69
500 000 lb/h	$ 12 M	$16 M	$0.99
2 500 000 lb/h	$ 50 M	$75 M	$0.83

Environmental Equipment

Plant Size	Equipment Cost	Installed Cost	Fixed Cost Per K lb/h Steam
100 000 lb/h	$ 0.8 M	$ 1.3 M	$0.36
500 000 lb/h	$ 3.5 M	$ 5.5 M	$0.30
2 500 000 lb/h	$15 M	$25 M	$0.28

Electric Plant

Plant Size	Equipment Cost	Installed Cost	Fixed Cost Per K lb/h Steam
s 5000 kW	$ 0.65 M	$ 1.05 M	$0.29
25 000 kW	$ 4 M	$ 6.3 M	$0.33
125 000 kW	$16.3 M	$19.5 M	$0.22

nomic feasibility for cogeneration is likely.

Two other major items for consideration are the value of electricity sold to the utility and Governmental regulations (such as PURPA — Public Utilities Regulation Policy Act). However, these two areas are beyond the scope of this text.

8.4 Size Considerations. The data in Table 32 illustrates that the economies of scale should have a decided effect upon the unit costs of larger coal-burning steam-producing facilities in fuel transportation and coal terminal facilities. This data also illustrates that there should be ample economic incentives to create large cogeneration facilities and thereby gain the advantages of lower unit costs per pound of steam.

A planning item which cannot be overlooked, is the need for redundancy in the plant. Redundancy can take the form of additional units, additional standby capacity, or both. The particular redundancy requirements of a given system could adversely affect the unit costs of a large facility to a greater extent than a smaller plant.

Redundancy requirements and other considerations indicate that small cogeneration plants will also be built. Industry requirements will mandate that they be built, even though they may not be as economical as larger plants. Certain site locations, individual plant systems, and fuel handling considerations will not permit larger facilities in some circumstances.

The data in Table 33 indicate that smaller cogeneration systems have a place in the planning effort. Smaller systems may be as economical as the very large systems through the use of less sophisticated and, therefore, less expensive equipment systems. Table 33 indicates specific economic data for a third party-owned cogeneration case for a smaller industrial plant requiring 200 000 lb/h of 125 psig saturated steam and 6500 kW of electrical power.

As a comparison, Fig 34, indicates the steam costs for a larger joint venture cogeneration project cf 1 125 000 lb/h of steam and 50 000 kW of electrical power.

It should be stressed that the eligibility requirements for cogeneration for specific plants will be highly variable, and each case should be considered on its own technical and economic merits. In some cases, legislative and regulatory requirements may override the technical and economic factors.

Table 33
Small Industrial Plant
Basic Economic Parameters

Item	Basis	Dollars
Facility cost	1981; installed cost	14 000 000
Annual op. cost	1981; including fuel	4 000 000
Unit fuel cost	1981 Base; per M Btu	1.90
Unit steam cost	1981; per 1000 lb/hr	4.25

(Unit steam cost includes credit for 6500 kW of electrical power) (39M kWh/y)

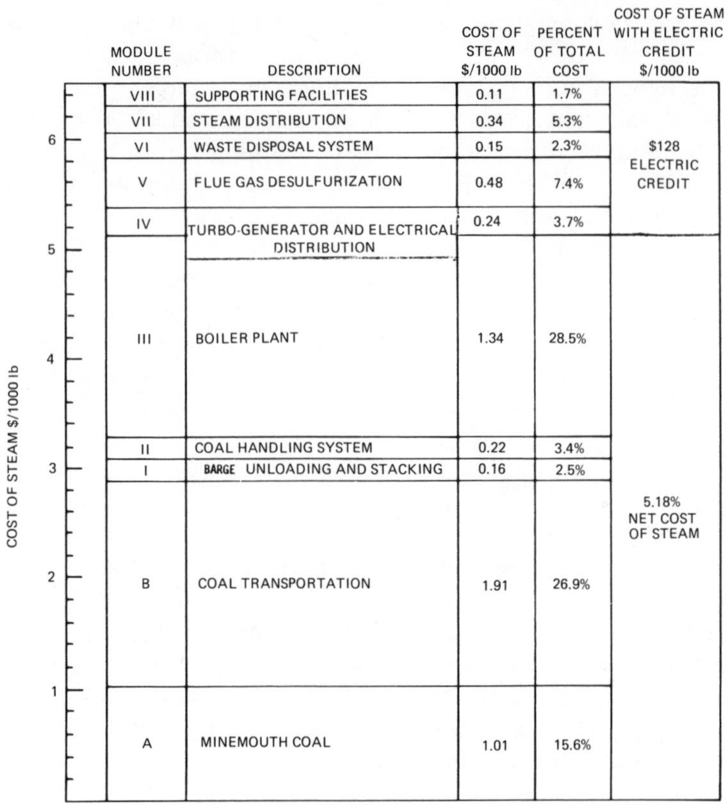

| | | | COST OF STEAM | PERCENT | COST OF STEAM WITH ELECTRIC CREDIT |
MODULE NUMBER	DESCRIPTION		COST OF STEAM $/1000 lb	OF TOTAL COST	$/1000 lb
VIII	SUPPORTING FACILITIES		0.11	1.7%	
VII	STEAM DISTRIBUTION		0.34	5.3%	
VI	WASTE DISPOSAL SYSTEM		0.15	2.3%	$128 ELECTRIC CREDIT
V	FLUE GAS DESULFURIZATION		0.48	7.4%	
IV	TURBO-GENERATOR AND ELECTRICAL DISTRIBUTION		0.24	3.7%	
III	BOILER PLANT		1.34	28.5%	
II	COAL HANDLING SYSTEM		0.22	3.4%	
I	BARGE UNLOADING AND STACKING		0.16	2.5%	5.18% NET COST OF STEAM
B	COAL TRANSPORTATION		1.91	26.9%	
A	MINEMOUTH COAL		1.01	15.6%	

Fig 34
Graphic Summary of Typical
1982 Average Cogeneration Steam Costs

8.5 Typical Systems. The cogeneration facility can be a joint venture. An industrial concern may find reason to include adjacent firms or the electric utility in the construction and operation of a cogenerator.

The industrial cogeneration system is in widespread use today and has been for decades in the petroleum, chemical, pulp and paper, and many other energy-intensive continuous process industries. The long time utilization of the back pressure and extraction steam turbine, and the gas turbine with waste heat recovery systems constitutes the bulk of many existing industrial cogeneration systems. The major impact on these systems today is the problem of continued fuel availability, the drastic increase in fuel costs, and increasing purchased power costs. In many cases, modifications for efficiency improvement are feasible.

Industrial-utility cogeneration systems usually offer the greatest flexibility in new applications. Normally the utility can provide for swings in power requirements, reliability in backup capac-

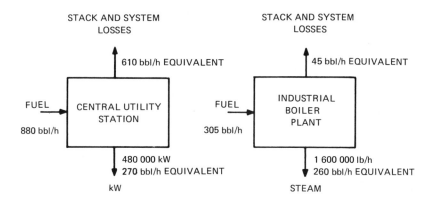

TRADITIONAL ARRANGEMENT

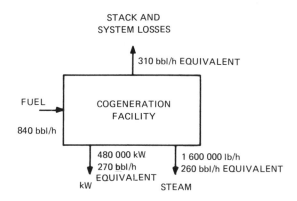

Fig 35
Cogeneration Fuel-Saving Potential

ity, and a *"sink"* for power generation in excess of the industrial's load. Thus, the basic system can be efficiently designed to match the process steam load (as a minimum) with the resultant kilowatts generated by the most efficient system. The advantage of this arrangement is diagrammed in the specific case as shown by Fig 35. This figure shows how a co-generation facility can produce the same end results in electric energy and steam production with a savings of approximately 30% in total fuel input. In equivalent oil barrels, while approximately 100 barrels of oil are required in the traditional arrangement, only 70 barrels are required in a cogeneration system (see Fig 36).

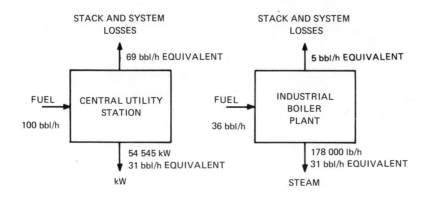

TRADITIONAL ARRANGEMENT

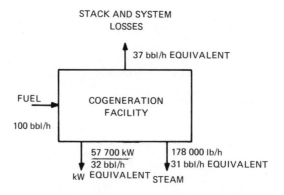

Fig 36
Cogeneration Fuel-Saving Potential
(Unit Comparison)

In Industrial-Industrial cogeneration, two or more relatively adjacent industrial plants form a coventure project so as to meet certain mutual requirements for process steam and kilowatts.

This arrangement may be particularly attractive when one plant requires relatively large kilowatt service and smaller steam requirements, and the other plant requires larger amounts of low-pressure

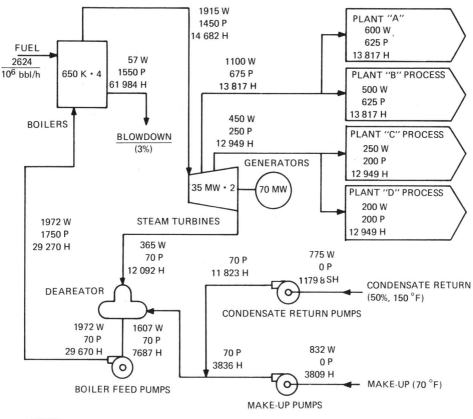

**Fig 37
Industrial-Industrial Cogeneration System**

process steam and a modest electrical demand. Figure 37 illustrates one such arrangement.

8.6 Other Considerations. One of the early requirements in developing a cogeneration system, especially when more than one party is involved, is the development of a business plan. This plan is essential for the generation of equitable and workable systems in the design, construction, operation, and management of a multi-owned cogeneration facility. A business plan provides for the initial management of the project, provides the vehicle for financial requirements, and provides a forum for the joint development of the project by all participants. One major consideration is the state or local public utility regulations. The joint venture will probably come under the scrutiny of various federal and state agencies. Should the economics appear favorable, legal analysis of regulations is in order.

Proper ownership shall be established. In many single-party projects the ownership alternatives are limited. In a multiparty project the ownership shall be decided on the basis of an equitable return to all, considering the favorable tax and credit incentives available. The

final decision will vary for each case and will seldom be the same for any two cases.

The principles for pricing, or prorated costing, of the steam and electricity produced from a cogeneration facility can become complex. The unit costs will vary from one system to another because of demand requirements, electric-steam mix, degree of participation in the project, and significant unusual requirements such as unusual pressure levels of steam required or very intensive electrical loads. One of the simplest ap-

**Fig 38
Steam /kW Cost Effect on
Product Energy Cost**

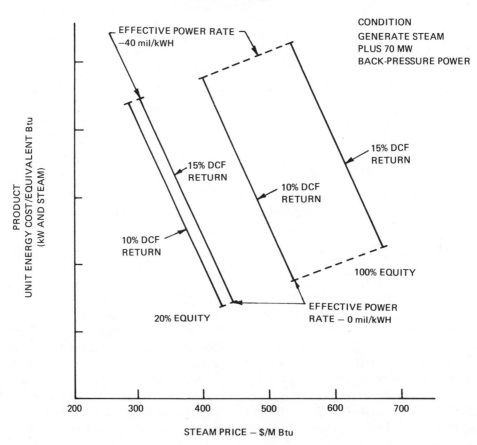

proaches to pricing steam and electricity is to equate both products to an equivalent Btu basis per unit of product. While simple in principle and in administration, this approach will not always be equitable to all participants. No single approach is feasible since each system will have varying mixes of product and financial participation. Furthermore, the distribution and energy extraction from steam cannot be measured as accurately as the distribution and use of electric energy. Figure 38 graphically demonstrates the effect on steam and kilowatt costs as various economic factors are held constant.

Some of the considerations that weigh heavily in the industrial planning effort are:

(1) Investment tax credits
(2) Conversion to coal credits
(3) Cogeneration credits
(4) Oil and gas use penalty taxes
(5) Exemptions from regulatory bodies
(6) Intertie regulations
(7) Tax life and depreciation schedules
These issues can establish a reasonable motivation for cogeneration facilities.

New or exotic technology is *not* required to make cogeneration a viable energy-saving approach. The equipment, systems knowledge and economic justifications exist today; and, except for the institutional barriers, energy-saving systems could be in widespread operation now.

In many instances, cogeneration will be the more practical and feasible approach as large scale conversions to solid fuels are implemented. Legislative incentives, environmental constraints, physical size restraints in existing plants, and financial requirements will provide the impetus for the development of energy-saving cogeneration facilities.

8.7 Bibliography

COMTOIS, WILFRED H. What Is the True Cost of Electric Power from a Cogeneration Plant?, *Combustion*, Sept 1978, pp 8-14.

HARKINS, H. L. Cogeneration for the 1980s, *Conference Record. IEEE Industry Applications Society Annual Meeting, 1978*, no 78CH1346-61A, 1978, pp 1161-1168.

JAVETSKI, JOHN, Cogeneration, *Power*, vol 122, no 4, April 1978.

SMITH, A. J. Remarks on Cogeneration, presented at October 25, 1977, American Society of Mechanical Engineers Conference.

STAFF REPORT. California PUC Staff Report on California Cogeneration Activities. *Resolution on Cogeneration* Jan 10, 1978.

Cogeneration, *Power Engineering*, vol 82, no 3, March 1978.

Index

Barge unloading and stocking, 144
Base period energy rate, 39
Baseboard heaters, 33
Basic economic parameters, 143
Basic energy balance, 33
Better utilization of equipment, 28
Biomass, 25
Blowers, 79
Boiler fuel, 27
Boiler plant, 144
Breakdowns, winding-to-frame in-
    sunation
  break-even analysis, 47
  bridges, 117
Bronze Book, 15
Building environmental shell, 33
Buildings
  commercial, 21
  public, 21

C

Capacitors, 78, 83, 116
  distortion, 83
  harmonic resonance, 83
    re-enforcement, 83
  increase in harmonics due to
    capacitor, 83
  thyristors, 83
Capital cost (investment), 43
  caulk, 31, 33
Certified ballast manufacturers
    (CBM) organization, 118
Chemical, 144
  potlines, 80
Chillers, 91
Chromaticity, 112
CIE color rendering index (Ra),
    115
Circuit, 122
  branch, 129
  solid-state control, 123
Closed-circuit television, 117
Coal, 123
  transportation, 144
  research, 23
Coal-burning facilities, 143
Coal terminal facilities, 143
Codes, 15, 18
Cogeneration, 22, 70, 137
  boiler feed pump, 139
  cooling towers, 137
  credits, 149
  determining the feasibility of,
    141
  electricity, 139
  electric power, 137
  entropy diagram, 139
  facilities, 143
  facility, 141, 145
  feasibility, approximations for de-
    termining, 141
  forms of, 137, 139
    bottoming cycles, 137
    compressors, 139
    electrical generator, 139
    fans, 139
    fuel saving potential, 145
    gas turbine-heat recovery boiler,
      139
    heat rejection losses, 141

high-pressure boilers, 139
high-pressure steam, 137
lakes, 137
ponds, 137
process steam, 137, 139
pumps, 139
rivers, 137
scale cost comparisons, 142
steam costs (average), 144
steam turbine, 139
systems, 137
temperature, 139
thermal energy, 137
topping cycles, 137
water conditions, 139
Color, 112
  rendering, 112
    index, 117
  temperature, 114, 117
Committee
  light (PAL), 15
Communication system effective, 29
Comprehensive audit, 30
Compressed air, 32
Compressors, 79
Compressor, centrifugal, 102
Conductors, 33, 76
  sizing, 65
Conservation
  concepts, 30
  considerations, 75
    in electrical machines and equip-
      ment, 75
  energy, 35
  methods of, 70
    organization for, 27
  opportunities, 30
  performance in plants, 35
  results, 38
Consolidation, 29
Construction Specification Institute
    (CSI), 21
Consumption patterns, 27
Contrast
  improvement, 110
  losses, 110
Conventional network equations, 81
Conveyor systems, 32
Cooling
  towers, 137
  water, 36
Current,
  distorted, 80, 81
  flow, 78
  generators of harmonics, 81
  imbalance, 76, 80
Cycle steam, combined, 139

D

Dampers, 31, 32
Data, 89
  kilovoltampere, 89
  kilowatt, 91
  power-factor, 95
Data collection, 29
dc resistances, 76
dc output voltage (average), 81
dc output voltage requirement, 82
Demand cost, 61, 62, 63, 64
Demand rate example, 59
Demand recorder, 95

Acknowledgment

Appreciation is expressed to the following companies and organizations for contributing the time of their employees to make possible the development of this text:

Advance Transformer Co
Celanese Fibers Co
Cleveland Electric Illuminating Co
Fiber Industries
General Electric Co
Moylan Engineering
North American Phillips Ltg Corp
Penn State University
Port Authority of New York and New Jersey
United States Department of Energy
Westinghouse Electric Co